The Romance of Aircraft

SEAPLANES NC-1, NC-3 AND NC-4 OF THE U. S. NAVY STARTING THE TRANS-ATLANTIC FLIGHT FROM ROCKAWAY

THE NC-4 ON ITS VICTORIOUS TRANS-ATLANTIC FLIGHT, SIXTY MILES AT SEA. THE SHADOW IS MADE BY A STRUT OF THE PHOTOGRAPHERS' PLANE

The Romance of Aircraft

The Development of Balloons, Dirigibles
and Powered Aircraft Including the
First World War in the Air

Laurence Yard Smith

LEONAUR

The Romance of Aircraft
The Development of Balloons, Dirigibles and Powered Aircraft Including the First World War in the Air
by Laurence Yard Smith

First published under the title
The Romance of Aircraft

Leonaur is an imprint of Oakpast Ltd

Copyright in this form © 2012 Oakpast Ltd

ISBN: 978-0-85706-992-4 (hardcover)
ISBN: 978-0-85706-993-1 (softcover)

http://www.leonaur.com

Publisher's Notes

Contents

PART 1

CHAPTER 1

The Conquest of the Air

On a beautiful afternoon in the latter part of the eighteenth century—June 5, 1793—a distinguished company of Frenchmen were gathered in the public square of the little village of Annonay, not far from Lyons. They had come there by special invitation of the brothers Stephen and Joseph Montgolfier, respected owners of a paper manufactory in the little town. It was whispered that the brothers had a great surprise in store for them, a remarkable discovery. Yet all their curious gaze could make out was a great linen bag, that swung, like a huge limp sail, from a rope that was suspended between two high poles. By means of this seemingly helpless piece of fabric the brothers Montgolfier proposed to accomplish the conquest of the air.

Those who ventured near to this strange object perceived at its base a wide circular opening, sewed fast to a wooden ring. The ring hung directly over a deep pit, in which had been heaped fuel for a bonfire,—straw and wood and chopped wool. At a given signal one of the brothers applied a torch to the mass and in an instant the flames shot up. A dense column of smoke arose through the neck of the bag. The latter gradually began to fill, spreading out in all directions, until, before the astonished gaze of the spectators, it assumed the shape of an enormous ball, that overshadowed the square, and that pulled and wrestled feverishly at the restraining ropes.

From the ranks of the onlookers a great shout of applause went up. The keepers let go the ropes, and the globe, like a live creature, freed from its bonds, rose triumphantly before their eyes. Up, up, higher and higher it went, so fast that they could scarcely follow it. For a moment it was hidden behind a patch of cloud, then it reappeared again, still ascending, until it rode majestically in the heavens, seven thousand feet above their heads!

The shouts and cries of the onlookers were deafening. Like wildfire the news spread from house to house of the little French village. Grave old legislators who had witnessed the surprising spectacle forgot their dignity and tossed their hats in air. Women, seeing the unusual object from a distance, fell on their knees to pray, thinking it a sign in the heavens, that portended, who knew what?

Man's age-old dream of conquering the air was now, for the first time, an accomplished fact. Those who stood in the little public square of Annonay on that auspicious afternoon long ago, watching the first Montgolfier globe on its victorious ascent, knew that it could be but a very short time indeed until men would be able to explore at will the dim regions of the upper air.

Meanwhile picture the consternation and terror among a group of humble peasants, who were tilling the fields a short distance from the spot where the famous Montgolfier balloon was launched. Suddenly in the sky there appeared a great black moon, which slowly and ominously descended toward the earth. The village priest himself led forth a little band of his stout-hearted followers to attack this dread instrument of the Evil One. With pitchforks and scythes they rushed upon the unfortunate balloon as it lighted gently on the ground, heaving this way and that with every puff of breeze that blew against it. With courage born of fear they prodded and beat the unfortunate monster. When the gas had finally escaped through the great gashes in its sides, and nothing remained but a disordered heap of tatters and shreds, the proud "conqueror of the skies" was tied fast to a horse's tail, and the terrified creature galloped off with it into the open country.

But the news of the Montgolfier brothers' discovery spread throughout the length and breadth of France and the civilized world. The French king ordered a special demonstration at Versailles, before himself and the royal family. On this occasion a wicker basket was swung from the richly ornamented balloon. In order to test the safety of travel in the skies there were placed in it a sheep, a cock and a duck. A fire was lit beneath the base of the balloon and it was filled with heated air. It rose with its strange cargo to a height of 1500 feet, traveled along peacefully two miles with the breeze and descended slowly into a near-by wood. There two gamekeepers, hurrying to the scene, were amazed to find its occupants calmly feeding, apparently unaffected by their voyage.

This incident gave the experimenters renewed courage and enthusiasm. A gallant Frenchman, Pilâtre de Rozier, volunteered to be the first man to make the ascent into the skies. A new and stronger machine was constructed, this time oval in shape instead of round, 74 feet high and 48 feet in diameter. At the bottom was a huge circular opening, 15 feet across. Just beneath this there was swung from iron chains an open grate, on which the fire was built by means of which the balloon was inflated. This grate hung down into a wicker basket or "gallery," in which the occupant stood, heaping fuel upon the fire. For of course, as soon as the fire died down, the heated air in the balloon commenced slowly to escape, and the whole thing sank to earth.

Pilâtre de Rozier was not at first permitted to set himself free and go voyaging unguarded into the upper air. Who knew whether this air above the clouds was fit to breathe?—who, for that matter, knew whether there actually *was* air at any distance above the surface of the earth? It was considered the better part of valor to try the experiment the first few times with the balloon tied firmly to the ground, with strong cables which only permitted it to rise some eighty or ninety feet. Even with these precautions a good deal of apprehension was felt regarding the healthfulness of the sport. But a sigh of relief was breathed by those who had the undertaking in charge when the bold de Rozier insisted that never in his life before had he known any experience so pleasurable as this of rising far above the housetops and of feeling himself floating, gently and peacefully, in a region of noiseless calm.

Impatient of this mild variety of aerial adventure, de Rozier finally won permission to make a "free" ascent, and he and his friend, the Marquis d'Arlandes, made a number of daring voyages in the Montgolfier fire balloon. Assuring their friends that no harmful results could come to them from ascending into the clouds, they loosed the ropes and went merrily sailing away until far out of sight. So long as they kept the fire in the grate burning the balloon remained aloft, and floated along in the direction in which the wind bore it. When they wished to descend they had merely to put out the fire, and as the heated air gradually escaped, the balloon sank gently to earth.

But the dangers of this sort of aerial adventure were very great indeed, and it required the most remarkable heroism on the part of de Rozier to undertake them. A chance spark from the grate might at any moment set fire to the body of the balloon, and bring it, a flaming firebrand, to earth. De Rozier understood this, and on his very first voyage carried along in the gallery of the balloon a bucket of water and a sponge. It was late in November of 1893 that he and d'Arlandes floated over Paris,—de Rozier heaping fuel upon the grate and tending the fire which kept the balloon afloat. Suddenly d'Arlandes heard a slight crackling noise high in the balloon. Looking up he caught a sight which turned him cold with horror,—a tiny licking flame far above his head. He seized the wet sponge and reached up to extinguish it. But another and yet another appeared, little tongues of fire, eating away at the body of the balloon. As each showed its face water was dashed upon it. From below the balloon could be seen peacefully journeying across the city. But far above, in its basket, de Rozier and d'Arlandes were coolly beating off the danger that hung over them like a Sword of Damocles. Not until they had been in the air twenty-five minutes, however, did they put out the fire in the grate and allow themselves to sink to earth.

These early experiments of the Montgolfiers and de Rozier fired the imaginations of scientific men in every part of the world, and it was only a very short time before a safer and more reliable type of balloon than the fire balloon was developed. Stephen Montgolfier's invention was based on the idea that smoke and clouds rise in the atmosphere. "If," said he to himself, "it were possible to surround a cloud with a bag which did not permit it to escape, then both would ascend." Of course this was a rather childish explanation of the cause of a balloon ascension, but it was the best that the Montgolfiers or any of their learned friends knew at that early day.

Now it was only a little while before this that an Englishman had discovered the gas which is now known as hydrogen, but which was then called "inflammable air." This gas, of which the Montgolfiers apparently knew nothing, is exceedingly light, and therefore rises very quickly in the air. The year before the Montgolfier balloon was invented, this Englishman, Cavallo, tried to fill small bags with hydrogen gas, on the theory that they would rise in the atmosphere. He failed merely because he did not hit upon the proper material of which to construct his bags. The fabric he chose was porous, and the gas escaped through it before the balloon could rise. Cavallo did, however, succeed in blowing hydrogen into ordinary soap bubbles, which arose with great velocity and burst as they struck the ceiling.

The problem of the material to be used in balloon construction had been fairly well solved by the Montgolfiers. Their balloons were of linen fabric, varnished and lined with paper, to render them as nearly as possible air-tight. This set the philosophers of Paris thinking how they might construct a globe which could be successfully inflated with hydrogen.

The brothers Roberts and M. Charles made the first hydrogen balloon. It was fashioned of very fine silk, varnished with a solution of gum elastic. This made it impossible for the hydrogen to leak through. The balloon was filled through an opening in the neck, which was fitted with a stopcock, so that the gas could be poured in or allowed to escape at will.

The funds for constructing this first hydrogen balloon had been raised by popular subscription, and the whole French people were alive with enthusiasm over the success of the experiment. Even at that early day France was the ardent champion of aerial conquest.

The day set for its ascension was the 27th of August, 1783. By the night of the 26th it had been partially filled with gas. It was tied to a cart, and long before daylight, started its journey to the Field of Mars. Throngs of spectators crowded every avenue. From the roof tops thousands of eager men, women and children peered down upon it through the darkness. Every window in every building was crowded with faces. A strong military guard surrounded it, riding on horseback and carrying flaring torches.

All day long multitudes crowded and jostled each other impatiently at the point where the ascension was to take place. At five o'clock in the afternoon the sudden booming of artillery fire gave notice to the hundred thousand waiting that the great event was on. Released from its bonds the balloon shot up, and in two minutes it was over 3,000 feet above the heads of the watchers. Still it continued steadily to rise, until finally it was lost to sight by the heavy storm clouds through which it had cut its passage.

The spectators were overjoyed, as on that first occasion when the Montgolfier balloon rose into the skies. It was pouring rain, but they did not seem to notice it as they cheered themselves hoarse at the second great air victory.

The balloon, likewise, was undiscouraged by the rain. Far above the clouds, where all was quiet sunshine, it journeyed peacefully along for fifteen miles, and descended in an open field.

The first two important chapters in the history of ballooning had now been written. Looking back, we are filled with gratitude to the French, whose courage, intelligence, and boundless enthusiasm made possible the conquest of the skies.

In other countries, of course, experiments were also in progress,

Montgolfier Experiment at Versailles, 1783

THE FIRST CROSS-CHANNEL TRIP

though they lacked to a great extent the popular backing which helped the French efforts to bear such splendid results. In London, an Italian, Count Zambeccari, constructed a hydrogen balloon of oil silk, 10 feet in diameter and *gilded*, so that in the air it was dazzling to look upon. A few months after the three Frenchmen launched their hydrogen balloon in Paris, this gorgeous affair was sent up in London, in the presence of thousands of spectators. One month later still, the city of Philadelphia witnessed the first ascension of a hydrogen balloon in the New World. It carried a carpenter, one James Wilcox, as passenger.

"What is the use of a balloon, anyway?" Benjamin Franklin was asked when in Paris at the time of the Montgolfier experiments. "What is the use of a baby?" the great American replied, smiling. Perhaps he had some inkling of the remarkable future in store for the science of aeronautics, then in its infancy!

The first really notable ascent in a hydrogen balloon after the early efforts was that of a Frenchman, M. Blanchard, who rose from Paris in 1784, accompanied by a Benedictine monk. Before they had got far above the ground a slight accident brought the balloon bumping down again. The monk, thoroughly scared, abandoned his seat, and M. Blanchard ascended alone. This balloon was fitted out with wings and a rudder, by which it was hoped to steer its course, but they proved useless, and its occupant had to allow himself to drift with the wind. He reached a height of 9600 feet, remaining in the air an hour and a quarter. Suffering from the extreme cold which is experienced so high in the atmosphere, and almost overcome with numbness and drowsiness, he was at length compelled to descend.

In England at about this time, Vincent Lunardi accomplished a free ascent in the presence of the Prince of Wales. But again it was the Frenchman, M. Blanchard, who succeeded in making the first *long* balloon voyage. In January, 1785, he and Dr. Jeffries, an American physician, sailed across the English Channel from Dover. It was a perilous adventure, with the ever present danger of falling into the sea. Half way across they found themselves descending. Then began a constant throwing out of ballast in a race with time and the wind. When the bags of sand they had brought for the purpose were exhausted they hurled overboard bottles, boxes, pieces of rope, even their compass and the apparatus of the balloon. They were still falling when in the distance they caught sight of the dim outline of the French coast, and in a last effort to keep afloat they began dropping articles of clothing over the basket's edge. Suddenly, however, the balloon began to mount. They floated in over the land, coming to earth safely not far from Calais.

Pilâtre de Rozier at once set about to imitate M. Blanchard's feat, and to avoid the danger of falling he constructed a hydrogen balloon with a fire balloon below it, so that by heaping on fuel he could force it to rise whenever he noticed a tendency to fall. In this ingenious contrivance he attempted to fly the Channel. At a height of 3,000 feet both balloons were seen to burst into flames, and de Rozier fell. So the gallant Frenchman who was first to explore the skies came to his unfortunate end.

His death cast a gloom over the many aeronautic enthusiasts of France, England and America. But his splendid pioneer exploits had borne their fruit in a permanent and growing interest in the navigation of the air. The science of aeronautics marched on, and new and important schemes were invented for conquering the skies.

CHAPTER 2

"A B C's" of a Balloon

Why does a balloon rise in the atmosphere?—is the very natural question we are apt to ask as we read the story of these early balloon experiments. The Montgolfier brothers themselves could probably not have answered it, for they claimed that some marvelous secret properties existed in "Montgolfier smoke." Stephen Montgolfier seems to have had the idea of "holding a cloud captive in a bag," since he had observed that clouds rise in the air.

The real explanation can best be understood by a simple experiment. Throw a stone into a pool of water and it will sink, because it is "heavier than water": that is, it weighs more in proportion to its volume than the same quantity of water weighs. But throw into the same pool a piece of cork and it will rise, because it is lighter in proportion to its volume than water. This truth was long ago expressed as a law by the old Greek philosopher Archimedes, who said:"*Every body immersed in a liquid loses part of its weight, or is acted upon by an upward force equal to the weight of the liquid it displaces.*" In the case of the cork, the weight of the water it displaces is greater than the weight of the cork, and consequently the upward force acting upon it is sufficient to lift it to the surface of the pool; but with the stone it is different: the water it displaces weighs *less* than the stone, and therefore the upward force acting upon it is not sufficient to prevent it from sinking.

Now all this applies just as well to a body in the atmosphere as it does to the body immersed in water. The air in this case corresponds to the liquid. Therefore any object placed in the air which weighs less in proportion to its volume than the atmosphere, is bound to rise. Every object we see about us, including ourselves, which is not fastened down to earth, would, if it were not "heavier than air," go flying off toward the skies.

Imagine a balloon all ready to be inflated, that is, ready to be filled with gas. The bag or "envelope" hangs limp and lifeless. Together with the basket, ropes, etc., which are attached to it, it probably weighs several hundred pounds, yet because its *volume* is so small it displaces very little air. Now we commence to inflate the balloon. As the gas rushes in, the envelope commences to swell; it grows larger and larger, displacing a greater volume of air every moment. When fully inflated it displaces a volume of air much greater in weight than itself. This weight of displaced air acts upon it with a resistless upward force, sufficient to lift it into the clouds. The moment its straining bonds are loosed, it rises with great velocity.

Of course, the lighter the gas that is used to inflate the balloon, the less weight will be added by it to the total weight of the structure,— although a lighter gas adds just as much to the volume as a heavier one would do. If two balloons of exactly the same weight before inflation are filled, one with the comparatively heavy coal gas which weighs ½ oz. per cubic foot, and the other with the very light hydrogen, which weighs 1/10 oz. per cubic foot, it is easy to see that the hydrogen-filled balloon will rise much faster and have a greater lifting power.

It is a simple matter to calculate what size balloon will be required to lift one, two or three passengers and a given weight of cargo, for we know that the balloon envelope must be large enough when filled with gas, to displace a greater weight of air than its own weight, together with the weight of the basket, equipment, passengers and cargo.

Once a balloon has been inflated and begun to ascend it would, if unchecked, continue rising indefinitely until it reached a point in the greatly rarefied upper air where it was exactly displacing its own weight, or, as science puts it, was "in equilibrium with the air." But this is usually not desirable, and so in all modern balloons arrangement is made for lessening the volume of the envelope and so decreasing the upward pressure. This is managed from the basket by pulling a cord which connects with a valve at the top and thus allows some of the gas to escape. There is also a valve in the neck of the balloon which opens automatically when the pressure becomes too great, or which can be operated by a cord. In addition to these two, balloons today have what is known as a "*ripping panel*," or long slit closed over with a sort of patch or strip of the envelope material. In case it becomes necessary to make a quick descent, the ripping panel may be torn open by pulling the cord which connects with this ripping strip. A wide rent is thus

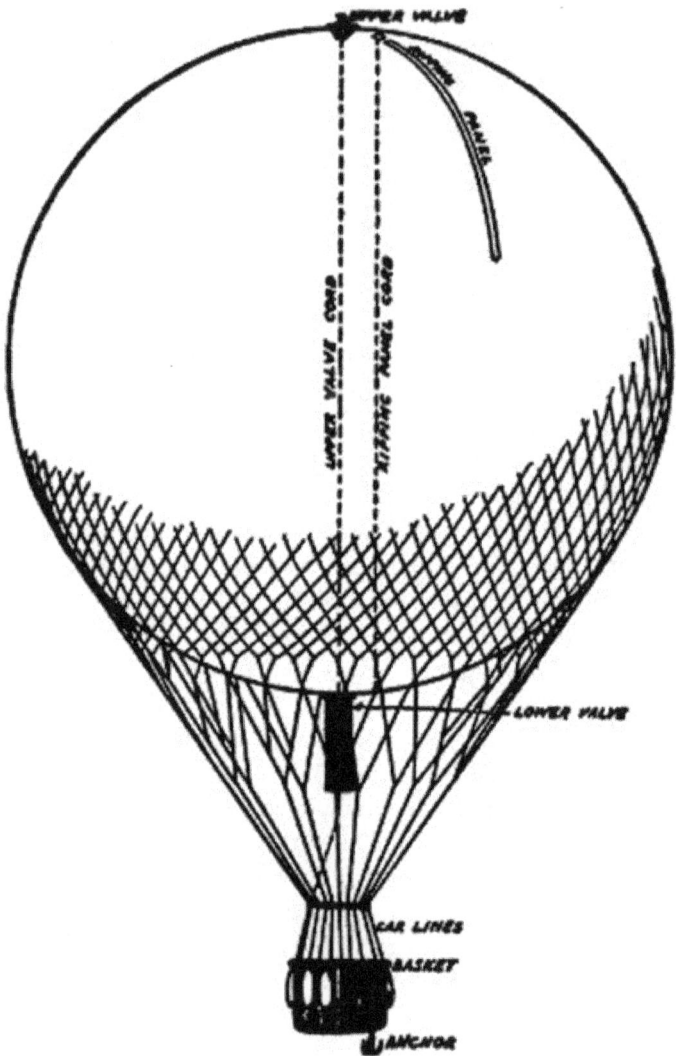

DIAGRAM SHOWING THE MAIN FEATURES
OF THE SPHERICAL BALLOON

produced in the envelope and the gas escapes very rapidly. As the balloon becomes deflated (that is, loses its gas), it grows smaller, displaces less and less air, and so sinks to the earth.

The accompanying diagram gives a very good idea of the main features of the spherical balloon. The envelope is usually made of strong cotton diagonal cloth, cut in pear shaped gores and varnished with a solution of rubber in order to prevent the gas from leaking through. At the bottom it ends in the long*neck*,—through this the balloon is inflated by joining it securely to a gas pipe which leads to the main supply of gas. Over the envelope there is spread a strong *net* made of heavy cord. From the net hang the stout *leading lines*. The leading lines in turn are attached to a strong wooden *hoop*, and from this hoop the car is suspended by ropes which are called *car lines*. The cords that connect with the upper and lower valves and the ripping panel hang down into the car and may be operated by the occupants, or crew.

Unless the balloon is held captive it is supplied also with a *trail rope*. This is a very heavy cable which is allowed to hang down from the car during an ascent. When descending, as the trail rope reaches the ground the balloon is relieved of a portion of its weight and becomes more buoyant. This makes its descent more gradual, for as it is relieved of one pound of weight of the dragging trail rope, it gains a slight tendency to rise again which counteracts the severity of its downward motion. The free balloon also has a *grapnel* or anchor for use in landing.

The *car* or *basket* of the balloon is usually made of woven willow and bamboo, which insures strength and lightness.

This brief description of the spherical balloon is intended to give the reader an idea of the essential features of any balloon. In modern warfare the captive balloon has proved its usefulness for purposes of observation, but the old spherical type is passing out. Balloons of many shapes and sizes, all designed for greater stability, are taking its place. Among these the"kite" or "sausage" balloon is by far the best known. Partly a kite and partly a balloon, with its long sausage-shaped body, its air-rudder or small steering ballonet attached to its stern, it possesses considerable "steadiness" in the air.

The kite balloon is used over the trenches to direct artillery fire and to report movements of the enemy: and it is likewise used over the sea, as a guide to direct the movements of the fleet in an attack, and as a sentinel on the lookout for enemy ships or submarines.

CHAPTER 3

Early Balloon Adventures

No sooner had the news of the remarkable balloon exploits of de Rozier and Blanchard spread throughout the nations, than people of all classes became interested in the future of ballooning. There were those who regarded it as the great coming sport, and there were also those who, like the French military authorities, saw in this new invention a possible weapon of war whose development they dared not neglect. It was only a short time before the French had an army training school for aeronauts, and a number of military service balloons.

The romance of ballooning had captured the imaginations of great masses of people and they proved their eagerness to back up the efforts of sportsmen balloonists with the necessary funds to carry on the many aeronautic projects which were suggested. We have already mentioned Chevalier Vincent Lunardi, the young Italian who was the first to accomplish a voyage in a balloon in England. The English people had read with ever increasing curiosity and impatience the stories of the French balloonists. What was their delight when this young Italian, poor but clever, proposed to give them an exhibition of their own. He had little difficulty in obtaining permission for a start to be made from London.

The next step was to obtain funds by popular subscription for the construction of the balloon. For a time money flowed freely into the coffers; but a Frenchman named Moret came into the limelight as a rival of Lunardi and announced a balloon ascent some little time before that planned by his opponent. The demonstration promised by Moret never came off, his balloon refused utterly to take to the air, and the indignant spectators went home, feeling that they had been cleverly hoodwinked out of the price of admission. Their wrath naturally turned upon the unfortunate Lunardi, and it was only with difficulty and after much discouragement that he actually succeeded in carrying his undertaking to completion. Finally, however, he had his balloon built. The king had withdrawn his permission for a flight from the grounds of the Chelsea hospital, but he succeeded in securing another starting place, and announced that he was ready to demonstrate what the balloon could do.

Vast crowds gathered to witness the spectacle. The balloon itself was gorgeous to behold. It looked like a mammoth Christmas-tree ball, of shining silk, in brilliant stripes of red and blue. It was filled with hydrogen gas, and as it gradually took form before their eyes, the people shouted with excitement and eagerness.

It was a pleasant September afternoon in the year 1784. When all was in readiness, Lunardi, no less eager and excited than the masses who had gathered to witness his exploit, climbed into the car. The cords were loosed and in a few moments the balloon, in its gala dress, was soaring far in the sky. Lunardi enjoyed his flight immensely. After traveling along without a mishap for a considerable time, he decided to come down, but once he had touched the earth he was seized by the desire to soar again. Putting out some of his ballast he allowed the balloon to arise once more into the sky.

Finally in the late afternoon he came to earth for the second time, landing in a field and greatly terrifying the simple country folk who were at work there. He was cold and hungry after his long journey in the rarefied upper air, but happy at the remarkable triumph he had achieved. Henceforth ballooning would not be regarded with derision and unbelief in England. The English nation was as wild with joy as the French had been at the early balloon ascents. Lunardi was lionized and became the favorite of the hour; his presence was demanded everywhere and he was royally entertained by the foremost people of the realm.

The British Isles became, from this time on, the scene of many a thrilling adventure with the balloon. It was only a few years later that Charles Green, the most famous of all the early English aeronauts, began his voyages in the *Great Nassau*, the balloon whose name is even today a tradition. In it he started out, one fall day in the year 1836, carrying provisions for a long voyage, but with no idea where the winds would carry him. The great balloon passed out over the British Channel and in again over the coast of France. Day faded into twilight and twilight into the blackness of night, but still it continued steadily on its way. Through the darkness Green and the friends who accompanied him in the large car of the balloon peered anxiously over the side, trying to guess where they were being blown. Finally after an all night ride, the dawn began to break, and in the morning the great balloon was brought to earth on German territory. Green had accomplished the longest balloon trip of his day. In the years that followed he made many voyages, but none that won for him more renown than this one.

Since the days when Blanchard accomplished the first trip across the British Channel, and the fearless de Rozier sought to imitate him, a number of aeronauts had made interesting voyages between France and England. One of the most adventurous was that of Mr. C. F. Pollock, in July, 1899. Accompanied by a friend, Mr. Pollock ascended early one afternoon, and after a picturesque and beautiful trip across the English countryside, sailed out over the sea. Behind them rose the white cliffs of the English coast, while before them gathering clouds hung like a curtain, through which they peered anxiously. Suddenly the balloon began to fall, and, fearful lest they should land in the rough waters of the channel, they began throwing overboard the sand which they had carried along as ballast. By means of this they succeeded in rising once more to a height of seven or eight thousand feet. It was early evening. Far below the sea had ceased to roar. They floated along in a realm of silence where nothing was visible through the veil of mist.

At length the veil was broken by the dim outline of the French coast. On and on they drifted yet seemed to draw no closer to it. There it remained, always ahead of them, tantalizing and provoking. Their ballast was almost gone, and they had unpleasant visions of landing in the water within view of their goal. So calmly and evenly did the balloon move forward that it was practically impossible for its occupants to tell whether it was moving at all. As they peered ahead uncertainly, searching the sea for a vessel by which they might gauge their progress, they felt themselves once more commencing to sink. In another few minutes the rest of the sand had been thrown overboard. There was nothing left with which to check the falling of the balloon, which surely and ominously continued.

The French coast was still many miles away. Almost in despair the two aeronauts cast about them for something which could be hurled over the side to lighten the weight of the balloon. As a last measure they decided upon the anchor. In another moment they had tossed it into the sea. Relieved of so great a weight the balloon shot up with lightning speed. The coast was drawing closer, but after its first swift ascent the balloon commenced to sink again and the aeronauts almost gave up hope of actually reaching shore. But just about eight o'clock they discovered to their great relief that the cliffs that marked the coast were below them. In another few minutes they had sailed in over the land. They opened the valve of the balloon and effected a descent in a field, where they were soon surrounded by an admiring circle of French peasants.

It was only about ten years after the pioneer voyages of de Rozier that the balloon was actually used on the battlefield, for in 1794 the French employed it against the Austrians at Mayence and at Charleroi. Under the fire of the Austrians who sought to prevent him from ascending, the French Captain Coutelle rose in an observation balloon at Mayence to a height of over a thousand feet. At that height he was beyond the range of the Austrian guns and could sit at ease watching their movements and preparations, at the same time dropping communications to the officers below. By his pluck he made possible a French victory, although the Austrians, much chagrined at their own lack of observation balloons, declared that this sort of warfare was unfair.

It may surprise Americans to know that balloons were used to good purpose for observation work in our own Civil War, and that they assisted the army of the North to keep an eye on the movements of Confederate troops around Richmond. They were once more employed by the French during the siege of Paris in 1870 and 1871, when 66 balloons left the city at various times, bearing messages, passengers, and flocks of carrier pigeons, which were used for delivering return messages. One plucky Frenchman dropped thousands of messages from his balloon upon the German soldiers, warning them of France's determination to fight to the bitter end. The incident reminds us somewhat of similar ones in the Great War, when the Allied aviators bombed the cities of Germany with proclamations.

The first notable employment of the balloon by the British Army occurred during the Boer War. During the siege of Ladysmith captive balloons were used to good purpose for observation and they were likewise made use of during a number of battles and under heavy fire. The French again employed them during the wars in Madagascar. Balloons had by the end of the nineteenth century become an important adjunct of every great army, and had proved themselves indispensable. Strange to relate they have never been driven from the field, and although we have today the swift dirigible and the still swifter airplane, there are certain military duties which they can perform best.

While the balloon was thus becoming a recognized instrument of war it was likewise gaining in favor among sportsmen. In all the great nations Aero Clubs were formed and races and contests began to be announced. In 1906 Gordon Bennett made the offer of a Challenge Cup for the longest trip by balloon. The contestants were to start from Paris. On September 30th, 1906, sixteen balloons arose from the Tuileries Gardens and started on their way. An American, Lieutenant Frank P. Lahm, carried off the cup, accomplishing a total distance of 401 miles and landing in Yorkshire.

The second race for the Gordon Bennett cup was held in America, and was won by a German. The third was held in 1908 in Germany. The winner, Colonel Schaeck, made a dangerous descent upon the sea near the coast of Norway, where he was rescued by a fishing boat. Several other contestants had perilous adventures. The American balloon *Conqueror* exploded in mid-air, much to the excitement of the thousands of spectators who had gathered to witness the start of the race. Instead of crashing to earth, however, as they had expected, it sank down gently, the upper part of the envelope forming a parachute. The aeronauts made an amusing landing on a housetop, little the worse for their sudden drop of several thousand feet. Another American balloon landed in the branches of a tree, while several of the remaining contestants came down in the sea and were rescued. On the whole it was a thoroughly exciting race, and the news of it aroused intense enthusiasm for the sport of ballooning in many lands.

Chapter 4

The Parachute

The story of the parachute is inevitably linked in memory with that of the balloon. Those who look back a few years can remember when exhibition balloons were in their heyday, and the sensation the parachutist used to create as he leapt from on high and came flying recklessly through the air. For a breathless moment or two the parachute remained folded, and when, finally, its umbrella-like form spread out protectingly above the hero, a thrill of relief ran through the anxious crowd of spectators.

In the early days of ballooning the parachute was looked on as a sort of life belt the aeronaut might don in the event of a serious accident to his craft in mid-air. Many experimenters gave their attention to developing it for this purpose; but when it was found that the balloonist actually needed no protection, since the balloon itself would "parachute" to earth after an explosion, interest in the matter waned.

Today the parachute has come once more into prominence because of the heroic work it performed in connection with the kite balloon and with the airplane in the war, and so our concern in it has revived. Many stories reached us from the front, of artillery spotters who jumped to safety when their observation balloons were unexpectedly attacked by enemy airplanes. It has actually become the "lifebelt of the air."

More often in the early days of ballooning it was a source of grave danger to the plucky aeronaut who sought to try it out and improve it, and its history includes the record of several sad accidents.

It was in the very year that the balloon was invented that a Frenchman, M. Le Normand began experimenting with a contrivance resembling an umbrella, with which he jumped from the branches of a tree, and sank gently to earth, the parachute saving him from injury.

Successful as his first attempt was it seems that he afterward lost his nerve, and later attempts were made with animals placed in a basket below the parachute and dropped to earth from a considerable height.

Blanchard, the famous balloonist, became interested in the idea of the parachute, and made a number of very interesting experiments. While making a public ascent in a balloon at Strasbourg, he dropped over the side of his balloon a dog with a parachute attached to him. The spectators were greatly pleased when the little creature came to earth quite unharmed, and public interest in the contrivance as a means of saving life was aroused.

In 1793 Blanchard himself undertook to make a parachute descent. He was not wholly successful, for before he reached the earth the apparatus gave way and he crashed down heavily, fortunately escaping with nothing worse than a broken leg. In spite of his injury he did not give up the idea of the parachute as a "life belt" for the aeronaut, and looked forward to the time when it should be so improved that it could be relied upon to bring the aeronaut to earth uninjured if any accident should make it necessary for him to escape from his balloon in mid-air.

However it was again a Frenchman, M. André Garnerin who accomplished the first descent by parachute from a great height without injury. His parachute was attached to a balloon. At a height of several thousand feet in the air, he freed himself and descended gradually, alighting gently upon the earth. That was in 1797 and five years later he gave a public demonstration of his parachute in England. This time he was not so successful, for his apparatus broke before he reached the ground and he received a number of injuries by his fall.

The parachute actually saved a life, however, in 1808, when the aeronaut R. Jordarki Kuparanto, whose balloon caught fire in mid-air during a demonstration at Warsaw, leapt over the side with his parachute and came to earth unharmed.

The parachute which Garnerin and the early aeronauts used in their experiments was fashioned to resemble an umbrella. As the aeronaut descended and the swift current of air caused by the fall rushed up under this canopy, it tended to hold it in the air much as the wind supports a kite, and thus the force of the descent was broken. In the year 1837 an Englishman named Cocking, who had been studying the principles of the parachute, came forward with an idea which differed greatly from this. The parachute he invented resembled an umbrella

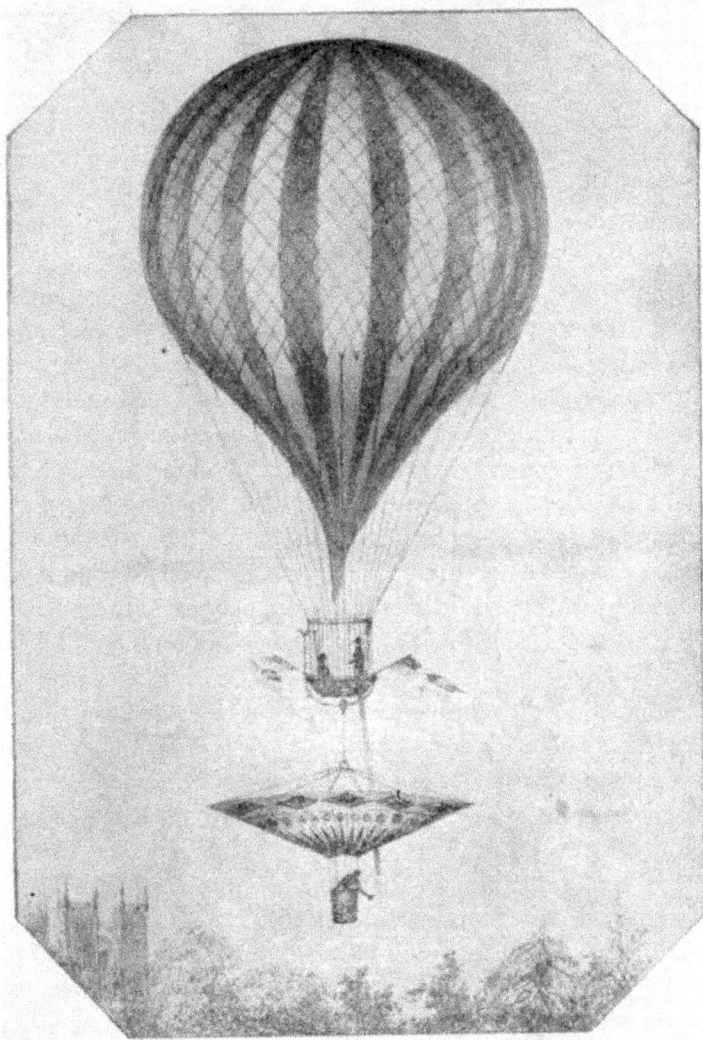

COCKING'S PARACHUTE

that had been blown inside out by the wind,—it was in other words an inverted cone, with a basket for the aeronaut hung from the cone's apex. The upper rim of the cone was made of tin to strengthen it, and the sides were of cloth.

Cocking was very enthusiastic over his invention, for he believed that his inverted parachute would descend more smoothly through the air than the old kind, which, while it supported the aviator, had a tendency to rock and pitch in the air after the manner of a kite. He sought an opportunity of giving his idea a public trial, but experienced aeronauts advised him not to do so, as they did not trust the safety of his apparatus. However, he insisted, and he finally persuaded the famous aeronaut Green to take him up.

On July 24th, 1837, the famous experiment was made. Green ascended in the great Nassau balloon, with Cocking's parachute suspended beneath it. Thousands of spectators had gathered to watch the ascent, but as the balloon was carried away by the breeze it was finally lost to their view, and so they were spared witnessing the accident which followed.

Green had been greatly worried over the safety of the parachute and had refused to free it from his balloon, but this difficulty Cocking had overcome by arranging a contrivance which permitted him to free himself when he thought the proper moment had arrived for his experiment. Finally, at a height of about 5000 feet, he called goodbye to Green and let himself go. Relieved of his weight the balloon bounded up with great swiftness, and it was some time before it recovered its equilibrium.

Meanwhile the parachute fell earthward with tremendous speed, rocking from side to side, until finally, unable to stand the strain any longer it went to pieces in the air, and the unfortunate parachutist came crashing to the ground. He died a few moments later.

Cocking's death cast a gloom over parachute enthusiasts, and for some time the contrivance fell into disfavor. But the real reason for its disuse was that balloonists found they needed no "life belt," as the balloon itself, if for any reason an explosion should occur, would sink gently to earth, the upper portion of the envelope forming a natural parachute.

So for a number of years the parachute was little heard of, except as a "thriller" at country fairs. In this connection it was always fairly popular. It was usually a folding umbrella parachute that the performer used on such occasions. As he leapt from the balloon he dropped

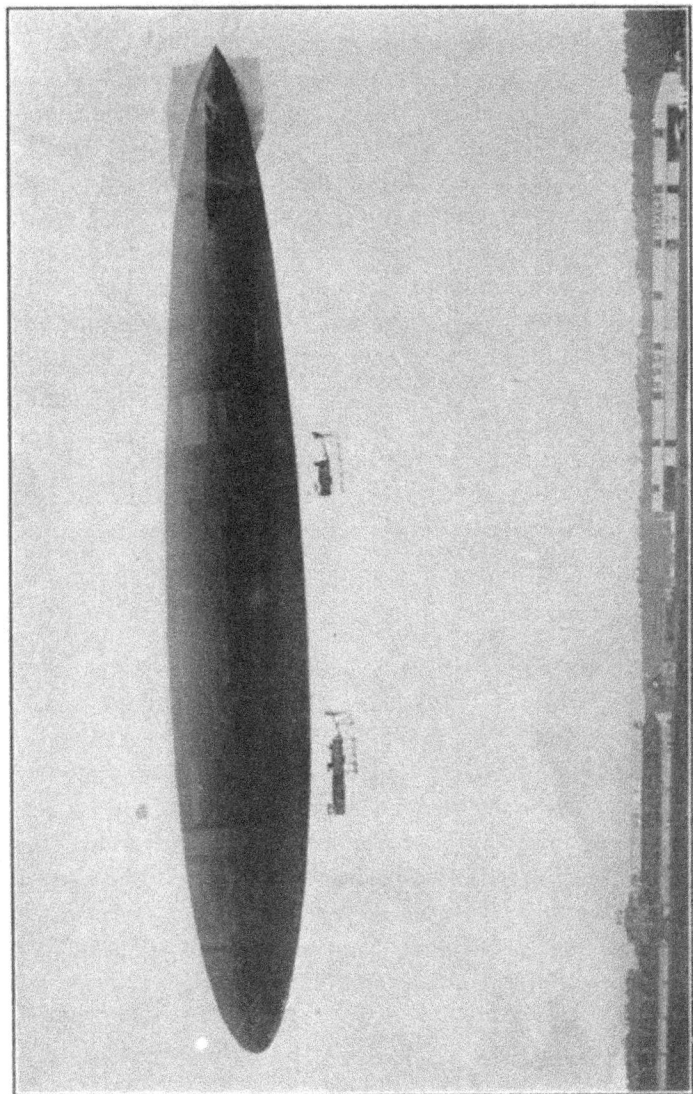

A GERMAN ZEPPELIN

straight down during a few terrifying seconds. Then to the relief of the spectators the parachute slowly and gracefully opened like a huge canopy over his head. From that moment his fall was checked and he sank gracefully and slowly to the earth.

With the coming of the Great War the day of the parachute was revived. Greatly improved in construction it came into its true and important *rôle* as the "life-belt" of the aeronaut. The life of the balloon observer in war times is a precarious one. His balloon is not free but is held captive by heavy cables reaching to the ground below. Hour after hour he sits watching the situation over the enemy's lines by means of a telescope. In the balloon basket he has a telephone which connects with the ground station, and thus he is able to send constant instructions to the artillery, enabling them to hit their objectives, as well as to keep the officers informed of the general situation. But his stationary position makes him an easy target for enemy bombs and bullets. At any moment he may find himself attacked by a squadron of airplanes. At the first indication of danger his comrades on the ground begin hauling his balloon down, and this precaution may possibly save his life. But often the emergency is very great. The aeronaut, attacked, unexpectedly and with no means of defending himself, has but one chance of saving his life, and that is to spring with his parachute from the balloon.

Thus the parachute was instrumental in saving many lives during the Great War, and in peace times it will probably be further developed for use in connection with the airplane as well as the balloon. Here the great difficulty lies in the fact that the pilot is strapped in his seat, and that he would not have time, in case of an accident in mid-air, to unstrap himself and attach a parachute device to his body. This might be overcome by having an apparatus already attached, so that all he would have to do would be to free himself from his seat and leap over the side. Here again he would run a very great danger of being instantly killed, as unless he manoeuvred his control levers just right before taking the leap, he would probably be hit by his own machine.

The idea has been suggested of a parachute arrangement to be attached to the upper wing of the airplane itself. This parachute would remain closed except in case of accident, when a lever operated by the pilot would cause it to open and carry the airplane safely to the ground. But the plan has never been worked out and it is impossible to say at this early date, (as at time of first publication), whether it would prove of much real benefit. In cases of engine failure the aviator can very often glide down safely to the earth; while in wartime, there is always the possibility that if the wings of the airplane were damaged by enemy fire the parachute also might be impaired.

An interesting use of the parachute was made by bombing airplanes and Zeppelins during the Great War. The pilots of these craft dropped flares or lights attached to parachutes, and by means of these they succeeded in locating their objectives and at the same time in "blinding" the operators of searchlights and anti-aircraft guns.

Just what the future of the parachute will be it is hard to predict. If there are to be future wars it will no doubt play an important part in them in the saving of human life.

The next few years will probably see the advent of huge aerial liners, built somewhat on the design of the Zeppelin. These great airships will travel in regular routes across the important countries of the world, bearing heavy cargoes of merchandise and large numbers of passengers. And we can easily imagine that in that day every traveller in the air will be supplied with a parachute as the ocean traveller of today is provided with a life-belt. Thus the simple little parachute will have performed its useful mission in the triumphal progress of aeronautics.

CHAPTER 5

Ballooning in the Great War

If you went down New York Bay during wartime you probably saw at the entrance of the harbour a United States cruiser stationed, with a "kite" balloon attached to it, standing sentinel against enemy submarines or aircraft. From their positions high in the basket, the observers could see far below the surface of the water, for the higher one rises in the air the clearer the depths of the water become to the vision. They had powerful glasses and by means of them could see far out over the water, where at any moment a periscope might have shown its face. The observers in that sentinel balloon could spot a submarine while it was still a long way off. A telephone connection reaching from the basket to the ship below made it possible for them to report a danger instantly, and soon the news would be travelling by wireless to the waiting destroyers and chasers.

This was probably the most important war duty that was being performed by a balloon on this side of the Atlantic. But over in Europe the kite balloon did valiant service above the trenches.

The coming of the heavier-than-air machine, with its powerful motor, its bird-like body, its great speed and lifting power, seemed at first to have driven the balloon from the field as an implement of war. And in fact, in the early days of the World War the airplane was almost exclusively employed by the Allies for scouting over the lines, watching enemy movements, directing artillery fire, and keeping the general staff informed of the strategic situation.

It was the Hun who first discovered that many of these duties could be far more efficiently performed by the "kite" or "sausage" balloon—the *drachen* balloon, as the Germans called it. This was not originally a German invention. It was first proposed in 1845 by an Englishman named Archibald Douglas, but his experiments did not meet with success and the undertaking was allowed to drop. Two Prussian officers, Major von Parseval and Captain von Sigsfeld, seizing upon the idea of the kite balloon as of great military importance, set themselves to developing it. In 1894 they produced the first *drachen* balloon, and it was this that gave the German army at the outbreak of the war one of its greatest advantages over the Allies.

The chief requirement for any observation balloon is that it shall rest in the air absolutely steady and motionless, so that the observer may not be interrupted in his study of the enemy's territory. The spherical balloon is apt to sway and roll with every puff of wind. The "kite" balloon therefore is a great improvement. Long and sausage-shaped, it combines the features of a kite and a balloon. Set at an angle to the wind, it is supported partly by the gas with which the main envelope is inflated, and partly by the action of the breeze blowing against its under surface, exactly as a kite is held in the air.

A smaller balloon, or steering ballonet, as it is called, is attached to the stern of the kite balloon and acts as a rudder. This ballonet is not inflated with the gas. It hangs limp while the balloon ascends, but the breeze quickly rushes into its open end beneath the main envelope and fills it out. This air-rudder, as it catches the breeze, acts as a steadier for the balloon. The main envelope has also an air chamber or section at the rear, which is partitioned off, and which is not filled with gas but is kept inflated by the action of the breeze; while on either side of the rudder there are two small rectangular sails, which help resist any motion of the breeze which might cause the balloon to sway.

Before the war the other large powers had made no attempts to imitate the German "*drachen*," although they had every opportunity of observing and studying it, and it seems very likely they actually underestimated its military importance. But when the war began, Germany surprised the Allies by the efficiency of these observation posts in the air. The fact that they were captive gave them certain advantages over the airplane for particular lines of work. They were able to direct artillery fire and keep the general staff informed of the situation over the lines. High in the air these lookouts could spot the tiniest change in the map. Provided with the finest instruments for observing, and connected with the artillery station or the headquarters by telephone, they could send in momently reports of the progress of the battle.

While the airplane was circling the sky to watch the effects of the last artillery fire, and had to get back to the ground before it could give full instructions to the gunners, the man in the basket of the kite balloon with a telephone in his hand, could report every second just where the last shell struck, whether the shooting was too high or too low, and how to vary the aim to get closer to the target. He was the eye of his battery.

The story of how the French military authorities at Chalais Meudon succeeded in obtaining plans for the first French military kite balloon was one of the carefully guarded secrets of the war. In the spring of 1915 the manufacture of kite balloons was well under way in France. In record time whole battalions of them were ready for service on land and on sea. They played a gallant *rôle* in the Dardanelles in connection with the British fleet. Soon afterward they were employed over the trenches in France.

The military kite balloon's first and chief aim is the directing of artillery fire. This it can do better than the airplane, which travels at high speed and must constantly circle or fly backward and forward in order to keep close to and be able to watch the target that is being aimed at. But the observer in the balloon basket sits practically motionless, while with the aid of a powerful telescope he watches the results of the firing. Before him he has a map on which he can plot the location of the target, and through a telephone connection he can advise the men in the ground station how to vary the range.

Think how much easier it is for him to explain to the men below by word of mouth the results of his observations, than for the observer in an airplane, soaring through the sky, to send that same message in a few brief words by means of wireless.

As a matter of fact the kite balloon at the front usually carries two observers in its basket: one to work directly with the artillery and the other to do general lookout work. The first has his eye on the target which the men below are trying to hit, and watches for the explosions of shells fired by his battery. But his comrade lets his gaze roam all over the horizon. He sees the movements of enemy troop trains, the massing of men and supplies, the flashes of the enemy's batteries. Should some objective of great importance loom up in the distance, such as a convoy of ammunition, the word is passed instantly to the battery below, and the guns are trained on it.

After the work in connection with the batteries, the second great *rôle* of the observation balloon is to keep the commanding officer at headquarters informed of the movements of the enemy, the effects of the firing and the general situation. The men in a balloon of this sort must know the territory very intimately, so that they can spot the tini-

Inflating a Service Balloon on the Field

est change. It is their duty to discover concealed batteries and other objects behind the enemy's lines which may help the divisional staff to lay its plans. And remember that they have no landmarks to go by. Out in that dread region of battle not a tree nor a mound has been left to vary the dull monotony of the brown earth, swept clean by the constant rain of shells. So it requires sharp eyes to distinguish the carefully camouflaged batteries of the enemy.

Of course the observation balloon at the front has to be carefully protected, for it furnishes a good target for the bombs from enemy aircraft. Every kite balloon has its detachment of defending airplanes, which circle round it in wide circles, on the lookout for approaching bombing planes of the enemy. Anti-aircraft guns also stand guard against the danger. Nevertheless the observer's life is a perilous one, the more so because he is a fixed target, unable to shift his position. A story is told of the heroism of Emile Dubonnet, the wealthy French sportsman, who was observing for the French "75's" near Berry-au-Bac when he was attacked by two German *taubes*. Appearing suddenly out of the clouds, they swooped down upon him, hovering over his balloon and dropping shells, which fortunately missed their aim. The *taubes* were so near to the balloon that the French were forced to stop firing lest they hit their own man.

Coolly Dubonnet continued his observations of the enemy's territory, telephoning the results of their fire to the French batteries below him, until a couple of French planes arrived on the scene and drove the *taubes* back to their lines. So severe is the strain of constant scanning of the enemy's territory through high powered glasses that it was found necessary to draw the observation balloon down about every two hours in order to change observers. At dawn the first balloons were sent up. All day long, except for the brief intervals when observers were changed, they stood there in the sky. Often far into the night they continued to play their silent *rôle* in the great drama of war. Some of the observers in fact became so experienced that they were able to do almost as good work at night as by day. It is said that enemy guns so camouflaged that they are not visible by day not infrequently show up in the darkness.

The kite balloon is connected with the earth by means of a strong steel cable, which winds onto an immense reel. To send the balloon up, the reel is turned and the cable is played out; when it is necessary to draw the balloon to earth once more, the cable is again wound about the reel. An electric motor is attached to the reel and turns it

36

Army balloon ready to ascend

in one direction or the other. Through the centre of the cable runs the telephone wire which connects the observer in the basket with the battery with which he works. The observer is equipped with a parachute for use in case of sudden danger. This parachute has straps like those of a man's suspenders which hold it to his back. When he springs from the balloon the parachute quickly opens and lands him gently and safely on the ground.

The kite balloon itself has been greatly improved since it was first constructed by the Germans. One of its greatest disadvantages lay in the great drag upon the cable, which when the wind was very high caused such an excessive strain that it was dangerous to use the balloon. The German "*drachen*" was badly "streamlined," that is to say, its shape offered great resistance to the wind. This resistance was increased by the rush of air into the open mouth of the steering ballonet.

An attempt to improve the design of the kite balloon was made by an American firm, the Goodyear Tire and Rubber Company of Akron, Ohio. They constructed a balloon which in general outline resembled the German "*drachen*," but which had not the steering ballonet or rudder at the stern. In its place they substituted large flat fins at the stern, and these, while they offered less resistance and thus reduced the strain or tug of the balloon upon its cable, did not hold the balloon absolutely steady in the air, as the steering ballonet had done. In order to give great steadiness the Goodyear people designed a tail like that of a kite, consisting of a number of very small inverted parachutes. These as they caught the breeze produced a resistance which steadied the balloon after the manner of the air rudder.

The Goodyear kite balloon was not an unqualified success, and it remained for Captain Cacquot of the French Army to produce the most satisfactory design. His was an almost perfect streamline model. Long and sausage-shaped like the German "*drachen*," it has, in place of the steering ballonet, three small ballonets at the stern which are in reality inflated fins. They are filled with air which is fed to them by a mouth or opening underneath the main envelope. These inflated fins, while acting as a rudder to hold the balloon steady in the air, do not offer the resistance that was caused by either the flat fins of the Goodyear model or the open-mouthed steering ballonet of the old type. Thus the French streamline balloon came to be the accepted model of the Allied nations, and proved itself an efficient arm of the service during the war.

Ballooning in itself will probably never be the sport that it once was, for the coming of the swift motor-driven dirigible and the still swifter airplane has made the old wind-driven vessel a hopelessly obsolete contrivance. It is therefore all the more interesting to know that the captive balloon, developed to highest form of efficiency, gave good service in the war against Germany and made itself a reliable and valuable servant of our armies, accomplishing its mission in a particular field in which neither the airship nor the airplane was able to compete with it successfully.

PART 2

CHAPTER 1

Development of the Dirigible

No sooner had the Montgolfiers and their colleagues constructed their earliest balloon models than scientific men and the general public, aroused by the possibilities of navigating the heavens, set themselves to devising schemes for steering aircraft. For of course the one great faculty which the balloon lacked was the ability to choose its own course. Once it arose into the air it was carried along in the direction and at the speed of whatever wind happened to be blowing.

Interest in the problem waxed so hot that there was scarcely a banker, farmer or grocer of those early days who did not have his private theory concerning the steering of balloons. Many learned essays on the subject were written, and many foolish solutions were advanced, among them that of harnessing a flock of birds to the balloon, with reins for guiding them. But the idea everyone thought most likely was that of oars, sails and a rudder.

Now there are several very good reasons why this method, adapted from sailing vessels, is useless when it comes to a balloon. In the first place, no sooner has the balloon risen to its maximum height into the atmosphere than it is caught in an air-current and carried along at exactly the same rate of speed as that at which the air itself is moving. To the occupants it seems to be hanging motionless in a dead calm, where there is no breeze blowing. Since its motion and that of the surrounding air are exactly equal, there is of course no resisting pressure against a sail, which simply hangs dead and lifeless.

To "row" in the air, on the other hand, would require oars of enormous size or else moving at a tremendous speed and a superhuman strength would be needed for moving them. Stop to think of the great velocity and power of the wind and then try to imagine the strength that would be necessary to row against this tide.

These facts, however, did not occur to the early experimenters, and balloons equipped with sails and oars were actually constructed. In order that they might present less resistance to the air, they were made egg-shaped, or long and cylindrical, sometimes with pointed ends, and this, at least, was an advance.

Another step in the right direction was the suggestion of paddle wheels, projecting from each side of the car, and beating the air as they revolve. This was coming very close to the correct solution, that of a revolving propeller.

But unfortunately at this early date the mechanical sciences were in their infancy, and although soon afterward the idea of a screw propeller did come up, the inventors were handicapped by the fact they knew of no other power than "hand-power" with which to drive it.

The man who might almost be called the father of the modern dirigible balloon was the French General Meusnier, an officer in the army and a man of great scientific and technical skill. Meusnier just proposed that air-bags or ballonets as they are now called be placed inside the balloon proper. By pumping air into these the balloon envelope could be filled out again when it had become partly deflated by loss of gas, for one of the great problems was to maintain the *shape* of the balloon after a quantity of gas had escaped. This was a good idea, but unfortunately its first public trial almost resulted in a tragedy. One Duke de Chartres ordered a balloon of this sort to be built for him by the brothers Robert, Parisian mechanics. Accompanied by the Roberts themselves and another man he ascended in it in July, 1784. The balloon was fish-shaped and was equipped with oars and a rudder. No sooner had it started on its upward journey than it was caught in a violent swirl of air which tore away the oars. The opening in the neck of the balloon became closed over by the air bag inside, and there was no outlet for the gas, which expanded as the balloon rose. Undoubtedly a terrific explosion would have occurred, but the duke, with great presence of mind, drew his sword and cut a slash ten feet long in the balloon envelope. He saved his own life and that of his comrades. The gas, escaping through the rent, allowed the balloon to settle slowly to earth, without injury to its occupants.

But the spectators did not understand the emergency, and the duke was covered with ridicule for his supposed cowardice.

The idea of the air-bags, however, was a useful one, and in later experiments worked well.

Meusnier gave a great deal of earnest study and experiment to the dirigible balloon, and he originated a design which was far ahead of his day. He decided on an elliptical or "egg" shape for the envelope, with small air bags inside it, and he suggested using a boat shaped car, which would offer less resistance to the air than the old round basket. The car was attached to the balloon by an absolutely rigid connection, so that it could not swing backward as the balloon drove ahead. Halfway between the car and the envelope he placed three propellers, and these, for want of any form of motor, were driven by hand pulleys.

Meusnier's design for a dirigible was the cleverest and most practical of its day, but owing to the cost, it was never actually carried out. In 1793, General Meusnier was killed at Mayence, fighting against the Prussians. After his death, little was heard of the dirigible balloon for another fifty years. Except perhaps for the novelty balloons at the country fair, the science of aeronautics slept.

The next appearance of the dirigible in history was in 1852, when the work of the Frenchman Giffard attracted widespread attention.

In 1851, Giffard had constructed a small steam engine, of about three horsepower, and weighing only 100 pounds. He thought it could be used for driving a balloon, and with the aid of a couple of friends he set to work building an airship, which was somewhat the shape of a cigar, pointed at the ends. It was 144 feet long and 40 feet in diameter at its thickest part, and it held 88,000 cubic feet of gas. Over the envelope was spread a net from which a heavy pole was suspended by ropes. At the end of this pole, or keel, as Giffard called it, was a triangular sail which acted as a rudder. Twenty feet below the pole hung the car, in which was the steam motor and propeller.

With this new means of driving the propeller, the dirigible began to show signs of proving a success, although as yet it could not develop any very great speed. One reason was that the engine was too heavy in proportion to the power it generated. Giffard's airship under the most favorable conditions could only go at from four to five miles an hour, when there was no wind.

One of the problems Giffard had to solve was that of preventing an explosion of the gas escaping through the neck of the balloon, as it came in contact with the heat of the engine. To avoid this, he placed a piece of wire gauze, similar to that used in safety lanterns, in front of the stokehole and the smoke of the furnace was allowed to escape through a chimney at one corner of the car, pointing downwards.

Giffard's second airship, of somewhat different design, was destroyed by an accident on its very first trip. He at once began working on a design for a giant airship, which was to be 1,970 feet long, and 98 feet in diameter at the middle. The motor was to weigh 30 tons, and he estimated that the airship would fly at 40 miles an hour. He worked out the scheme in every detail, but owing to the expense the dirigible was never made.

The first "military dirigible" ever built was that constructed by Dupuy de Lôme for the French government during the siege of Paris, and tried out in 1872. Its propeller was driven by a crew of eight men, a very curious proceeding, since the steam engine had been successfully tried.

A dirigible which was almost modern in design was meanwhile being constructed by Paul Haenlein in Germany, and made its appearance in 1872. It was long and cylindrical, with pointed ends, the car placed close to the balloon envelope, to give a very rigid connection. Its really noteworthy feature was the gas engine, replacing the steam engine that Giffard had used as a means of driving the propeller. The gas for the engine was taken from the balloon itself and the loss was made good by pumping air into the air-bags. The balloon envelope held 85,000 cubic feet of gas, and of this the engine consumed 250 cubic feet an hour. This dirigible, on trial trips, attained a very fair speed, which would have been greater had hydrogen gas been used in the envelope instead of ordinary gas. But lack of funds prevented further experiment, and Haenlein had to abandon his attempts.

Ten years now passed before the next notable effort at dirigible construction. The delay was probably due to the fact that no suitable driving power was yet known. In 1882 the famous French aeronauts Gaston and Albert Tissandier constructed an airship somewhat similar to Giffard's models, but containing an electric motor. But although this dirigible cost £2,000 or almost $10,000 to build, it had the same fault as all that preceded it; it could not develop speed. The problem of finding an engine of sufficiently light weight and high power was a difficult one, which has not today been wholly solved.

The public generally had begun to think of the dirigible balloon as impractical and impossible, when in 1884 came the startling news that two French officers, named Renard and Krebs, had performed some remarkable feats in a balloon of their own design. An electric motor of 8½ horsepower drove the propeller.

Several details of this dirigible are extremely interesting. The axis on which the propeller blades were fixed could be lifted in order to

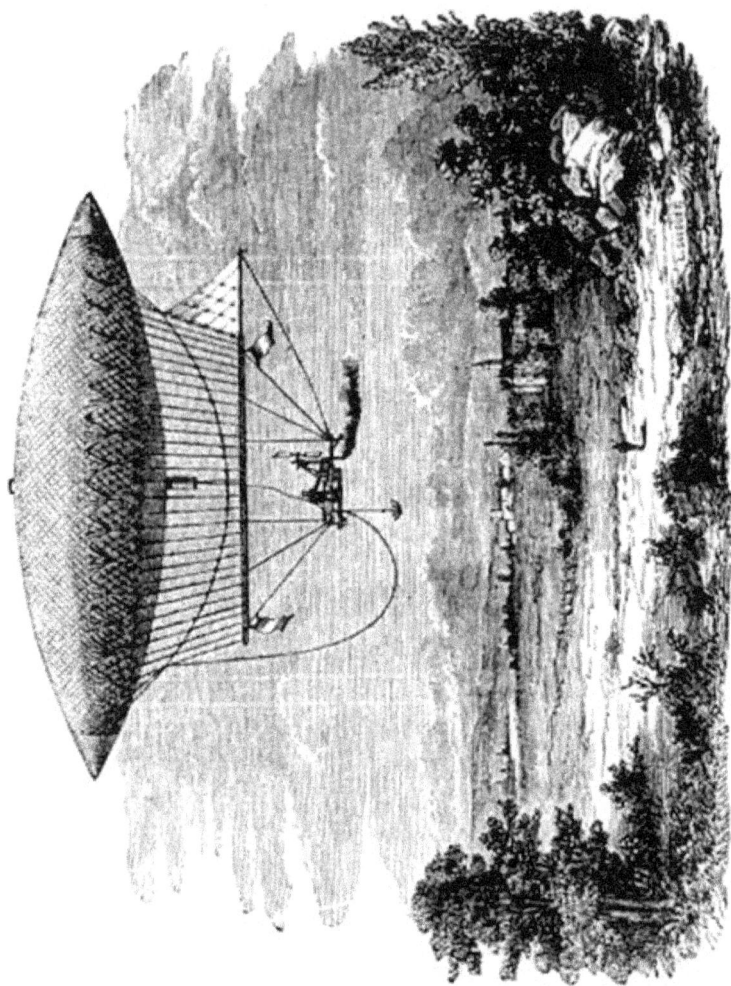

GIFFARD'S AIRSHIP

prevent them from being injured in case of a sudden drop. A trail rope was also used so as to break the shock which might result from a sudden fall. At the back between the car and the balloon was fixed the rudder, of unusual design, consisting of two four-sided pyramids with their bases placed together.

Renard and Krebs christened their dirigible "La France," and on August 9, 1884, they gave it its first public try out near Chalais, with great success. They travelled some distance against the wind, turned and came back covering a distance of about 5 miles in 23 minutes. Never before had a balloon been able to make a trip and return to the place of its ascension.

But in spite of the success of Renard and his comrade, construction of dirigibles in France paused for some time, and it was in Germany that the next attempts were made.

In 1880, a cigar-shaped dirigible, equipped with a benzine motor was demonstrated in Leipsic. It had been built the year before by Baumgarten and Wölfert. At its sides it had "wings" or sails and three cars were suspended from it instead of one. This airship met with a serious accident on its very first trip. A passenger in one of the cars destroyed the balance, the whole thing toppled over and crashed to the earth, the occupants miraculously escaping injury.

Not long afterward Baumgarten died. Wölfert constructed a new dirigible of his own design containing a benzine motor in which he ascended from the Tempelhofer Feld, near Berlin, in June, 1897. Wölfert had neglected to provide against contact of the gas escaping from the envelope with the heated fumes from the engine. An explosion took place in mid-air, and the machine fell to earth in a mass of flames, killing Wölfert and the other occupant.

Next in the long series of attempts came that of an Austrian named David Schwartz, who designed a dirigible with one entirely new feature: a rigid aluminum envelope. This balloon had a petrol engine. It was tried out in Berlin in 1897, but an accident to the propellers brought it crashing to the ground. Its occupant jumped for his life and barely escaped killing.

Up to this time there is little to record in dirigible history but a long series of valiant attempts and failures, punctuated all too frequently by gruesome disasters. But the nineteenth century was drawing to a close, the twentieth century with its era of mechanical triumphs was at hand, and the time was ripe for those champions of the dirigible to appear who should make it a potent factor in modern warfare.

SANTOS-DUMONT ROUNDING THE EIFFEL TOWER

Almost at the same time there stepped into the limelight of public interest two men, representing Germany and France, whose names are now famous in the aeronautic world. In 1898 there appeared in Paris a young Brazilian named Santos-Dumont, who began constructing a series of dirigibles whose success astounded the authorities.

In exactly the same year Count von Zeppelin, in Germany, formed a limited liability company for the purpose of raising funds for airship construction. His first dirigible balloon was the longest and biggest that had ever been built. Although the envelope was not, like Schwartz's dirigible, of solid aluminum, it was practically rigid, for it was made by stretching a linen and silk covering over an aluminum framework.

Zeppelin's first airship had two cars, with a motor in each, giving about 30 horsepower. On its trial trips it made a better speed than had yet been attained.

With the experience he had gained Zeppelin set to work on a new design. It was five years before he secured enough funds for its construction, but it was finally ready in 1905. The most important improvement was in the motors, which were as light in weight as those of the first dirigible but had a greatly increased power. As before, there were two cars, with an 80 horsepower motor in each.

Even this airship, in spite of its greater speed, was not an unqualified success, for it was discovered that it had too great a lifting power, so that when launched it rose at once to a height of about 1500 feet, and was impossible to operate at a lower level.

Santos-Dumont, meanwhile, in Paris, had been performing feats of aeronautics which had made him the acknowledged "hero of the air." Santos-Dumont was probably far from being the scientific student of balloon construction that Zeppelin was, but while his dirigibles did not attain a great speed or represent a tremendous advance in actual theory, his public performances served one great purpose, they aroused the ardour and enthusiasm of the whole French people and of many in other countries for the sport of ballooning. Santos-Dumont had great wealth, and a sportsman's courage. He constructed in all 14 dirigibles, each time seizing upon the experience he had gained and incorporating it into a new model, casting aside the old.

Santos-Dumont's airships were altogether different from those of Zeppelin. While Zeppelin's had an inner framework to maintain the shape of the envelope, Santos-Dumont depended entirely on the linen air bags, placed inside the balloon, which as it became flabby through loss of gas, could be pumped full of air to hold the envelope in place. His balloons were either long and cylindrical with pointed ends, "cigar-shaped," or else "egg-shaped," with ends rounded.

In spite of all the curious accidents that beset this young Brazilian on his early trips, in the vicinity of Paris, he was never once deterred from his efforts. He almost lost his life several times in his first airship, but he profited by the mistakes of construction in building the second. His dirigibles increased in size as he installed in each successive model a more powerful and consequently heavier motor, requiring greater lifting power.

In his third balloon Santos-Dumont ascended from the Champ de Mars in Paris and circled the Eiffel Tower amid the cheers of thousands of onlookers, finally descending in an open field outside Paris.

Public interest was now thoroughly aroused. A prize of £4,000 was offered by Monsieur Deutsch to the aeronaut who could circle the Eiffel Tower and return to the starting-point at Saint Cloud within half an hour. Santos-Dumont attempted this with his 4th and 5th machines, but it was not until he built his 6th model that he finally accomplished it. The Brazilian government sent him a gold medal and an additional £5,000 with which to build new balloons.

Number 9 was the most popular of all Santos-Dumont's machines. He became the idol of the French public, whom he was always surprising with his spectacular and unlooked-for adventures. During the races at Longchamps he descended on the race course, stayed to view the performance, then mounted in his car and rode away. He amazed the passersby by alighting before his own front door in Paris where he left his airship while he went and ate breakfast. He sailed up opposite the grandstand when President Loubet was reviewing the French troops, fired a salute, and as unexpectedly departed.

Santos-Dumont's power of escape from death seems almost uncanny but it was due to his coolness in facing any situation. In the majority of his airships he used a petroleum motor, and with this there is considerable danger of the petroleum in the reservoir catching fire. On one occasion a fire did start, but he succeeded in extinguishing it with his panama hat. Among all his mishaps, including that of falling into the Mediterranean Sea, he never really had a serious explosion.

Another young Brazilian, however, named Severo, was killed in a dirigible of his own construction, when the petroleum in the engine caught fire. He ascended in May, 1902, in a balloon which he called the *Pax*. His car was seen suddenly to burst in flames, a violent explosion followed, and the whole thing crashed to earth.

Santos-Dumont placed his last three dirigibles at the disposal of the French military authorities. Actually he had not developed a type suitable for military use. But his public performances had aroused intense popular interest and had succeeded in opening the eyes of the French authorities to the possibilities of the airship in time of war. His remarkable aerial feats had attracted the attention in particular of two Frenchmen of his own fine metal and courage, who from this time forth left no stone unturned to excel him in his achievements.

CHAPTER 2

Forerunners of the Allied Dirigibles

It is to the two French brothers Lebaudy that France and the Allies owe the credit for the development of the big military dirigible such as is used, (as at time of first publication), in the present war. These brothers were wealthy and full of enthusiasm for aeronautics. From a distance they had watched the achievements of Santos-Dumont and they determined to expend every possible effort to excel him in the construction of dirigibles. In 1899 they commissioned an experienced engineer named Jouillot to make a study of the problem, to discover if possible why previous experimenters had failed to produce a model of satisfactory speed and power, and to draw up designs for an airship which should correct the faults of those already known.

It took two years before a finger could be lifted toward the actual building, but finally in 1901 the work of constructing the first Lebaudy airship commenced. It was ready for a try out in November, 1902. The envelope was of bright yellow calico: it was cigar-shaped, 187 feet long and 32 feet in diameter. The envelope was fastened at the bottom to a rigid floor-work of steel tubing and from this the car was suspended. The dirigible was fitted with a 40 horse power benzine motor; and its total weight, including a supply of benzine, water and ballast, was two and one-half tons.

During the next year this dirigible made at least 30 trips, at very fair speed. Meanwhile the builders were studying it in every detail, working out ideas for improvements and drawing up plans for their next model. In 1904 they built their second airship. It was somewhat longer than the first and about the same shape, but the pointed end at the rear had been rounded off. Calico was again used for the covering of the envelope, and it was made absolutely air-tight by coating it inside and out with rubber. Besides the main valve there were safety valves in the envelope for allowing the gas to escape when the pressure became too great. The envelope was also provided with two small windows, so that the inside of the balloon could be easily inspected. It had sails to give it greater stability, and two movable sail-like rudders, placed together at a V-shaped angle. The driver could alter the position of the sails and the rudder according to the wind.

The car of this Lebaudy airship was boat-shaped with a flat bottom. To diminish the shock in case of a fall steel tubing was placed in a slanting position beneath it in a pyramid arrangement, the point facing downwards. The car was set very close to the envelope or body of the airship, and carried the 40 horse power benzine engine. At the front of the car was an electrically worked camera, a 1,000,000 candle power acetylene projector providing lighting by night.

Many improvements were later added to this second dirigible which was christened the *Lebaudy*. The interest of the French Minister of War was aroused and he appointed a commission from the Balloon Corps to follow the progress of the experiments.

Every one now began to look upon the dirigible as a factor to be reckoned with in the event of a war. The Lebaudy brothers offered their airship to the French government, and after it had accomplished a series of tests to prove its value as an instrument of war, it was accepted, and became a model for later airship construction.

Germany was not far behind, for already Count von Zeppelin's second airship had proved itself a success, and plans were being laid for a third. From this time on the two European nations destined to become powerful adversaries in the World War, though working along somewhat different lines, kept almost neck and neck in their struggle for air supremacy.

The French military balloon department began at once the work of constructing an air fleet with the *Lebaudy* as a model and with the engineer Jouillot as chief adviser, this work went forward with great rapidity. The *Lebaudy* was followed in design pretty closely, but a few changes were made which experience had suggested. For one thing the balloon envelope was rounded at the front and pointed at the rear, exactly the reverse of the Lebaudy model, as this arrangement was thought to offer less resistance to the air. It had an internal air-bag or ballonet whose capacity was one-fifth that of the envelope. This ballonet was of course empty on the ascent. It was calculated that the balloon could reach a height of about a mile. To descend, gas would then be allowed to escape, and, in order to keep the envelope fully inflated, air would be pumped into the ballonet.

This first type of dirigible actually constructed by the French Army was called the *Patrie*. It was 197 feet long and carried a benzine motor of from 30 to 40 horse power, which drove the two double-bladed steel propellers. As in the case of the *Lebaudy*, the *Patrie* was protected from injury by a strong steel framework, coming to a point below the car. In case of a sudden drop, this point would strike the ground first and ward off the blow from the car, and the propellers. Good as this plan *seemed*, it did not always work. The *Patrie*, after many successful journeys, met with an accident to her motor, escaped her guard of soldiers and drifted off alone. She crossed the English Channel and fell in Ireland, breaking off her propeller. Before she could be captured she rose again into the air, drifted out over the sea and was never again heard from.

M. Deutsch, who had done so much to encourage the efforts of Santos-Dumont, stepped forward in the emergency and offered the French government his airship the *Ville de Paris*. This had been designed for him by an engineer named Tatin. It was 200 feet long, made of German Continental Rubber Fabric, and, like the *Patrie*, had an internal air-bag of one-fifth its capacity. In one important respect it was different from those that preceded it. At its stern it had eight small cylinders, or ballonets, filled with gas, which added greatly to its stability, though they detracted from its speed by causing a considerable resistance to the air.

While the car of the *Patrie* was about 16 feet long, this new airship had a car measuring 115 feet, and the propeller was at the *front*, so that as it revolved it *drew* rather than *pushed* the car through the air. A propeller of this sort is termed a "tractor," and figures today in many models of aircraft.

During these years of experiment in France, England and America had looked on in comparative idleness. In 1902 England did indeed possess one small airship, designed by Colonel Templer of the Army Balloon Department, and christened the *Nulli Secundus* (*Second to None*). She was "sausage shaped:" rounded at the front and pointed at the stern with a peculiar rudder design. Her car was boat-shaped and her propellers were aluminum, both revolving in the same direction, which gave her a curious tendency to "somersault." In spite of their "baby" dirigible's rather pretentious title, the military authorities, and the English public in general, evidently took slight store in the infant prodigy, for from 1902 to 1908, she only came out of her shed for a few short trips. In 1908 she was completely remodelled, and emerged for a trial trip. But neither the government nor the public seemed interested in Colonel Templer's schemes. The valiant little pioneer ship of England's air fleet went back to her sheds, resigning herself to obscurity.

Our own country, which in many other lines has led the world in its mechanical skill and enterprise, did not have a single army dirigible till as late as 1908, when it gave out a contract for an airship which was built by Captain Thomas S. Baldwin. The motor was designed and built by a young mechanic in Hammondsport, N.Y., who for several years had been manufacturing motors for automobiles. His name was Glenn Curtiss and he afterward became one of the world's most famous aviators.

United States Army Dirigible No. 1 was long and cylindrical, pointed at both ends, and covered with Japanese silk, vulcanized with rubber. The water-cooled Curtiss motor was a 20 horse power, and the wooden propeller was of the "tractor" type, placed in the front of the car.

Germany, while America and England stood idle, had been rapidly forging ahead. By 1908 Count von Zeppelin had constructed his third and fourth models, and his public demonstrations had aroused the whole German people to unbounded enthusiasm. The Crown Prince made a trip in Zeppelin No. 3 and its originator was decorated with the Order of the Black Eagle. The German Association for an Aerial Fleet was formed, and within a short time over a million dollars had been contributed by the people for the purpose of building dirigibles.

Zeppelin No. 4 was destroyed by an accident, but Zeppelin No. 3 was recalled into the national service and in 1909 given the official title of *S.M.S. Zeppelin I*. From this time on dirigible construction in Germany went forward with the greatest speed. Two other names became prominent in the enterprise: those of Major von Parseval and Major von Gross. The "Parseval" design resembled more the French, for it was covered with "Continental fabric," was long and cylindrical, rounded at the front and pointed at the stern, with a large internal air ballonet. The car was suspended from two steel cables or trolleys, which it could slide along, altering its position and the "balance" of the whole airship.

The "Gross" type of airship resembled the *Lebaudy* and the *Patrie*, with its boat-shaped car hung from a steel platform attached to the

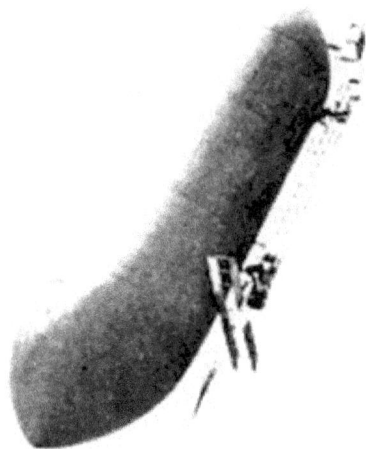

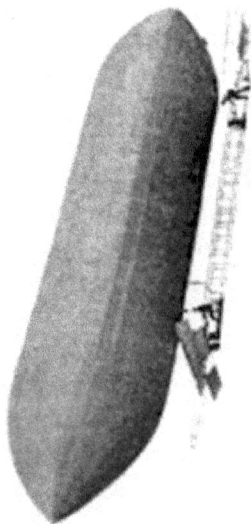

BALDWIN U. S. "DIRIGIBLE NO. 1"

bottom of the envelope.

Out of this brief story of the development of the early airship models of all the nations, we can, if we look carefully, see certain definite types of dirigibles emerging. The experimenters had to solve this problem: What shall we do when owing to loss of gas the balloon envelope begins to get flabby? For of course a flabby, partially filled envelope would flop from side to side, destroying the balance of the airship and checking its speed.

The German inventors settled the problem by making the envelope *rigid*, either with a solid covering or with a covering of fabric stretched over an inner framework. Thus the *rigid type* of airship was evolved.

The French inventors solved the same problem by placing inside the envelope a large *empty* bag of fabric, into which air could be pumped when necessary to fill the balloon out and hold the envelope firm. The air could not be pumped directly into the envelope itself as it would produce an explosive mixture with the gas already there. From this method of dealing with difficulty, the *non-rigid* type of dirigible was evolved.

But the *non-rigid* dirigible presented a new difficulty: how could the car be suspended from it in such a way that it would not swing? For only with a rigid connection between the car and the envelope could the greatest speed be obtained. The *Lebaudy* solved this problem by attaching to the base of the envelope a rigid steel flooring, from which the car could then be suspended by an immovable connection. And so was evolved the *semi-rigid* type of airship.

In recent years another solution of this problem of preventing the car from swinging has been employed to some extent: By making the car almost as long as the envelope, the connecting cables by which the car is suspended hang almost perpendicular, and there is not the same tendency to swerve as with cables slanting down to a comparatively small car. This type of airship is called the *demi-semi-rigid*.

These then are the four general classes of dirigibles which were used in the Great War.

THE BRITISH ARMY "BABY" DIRIGIBLE

CHAPTER 3

Dirigibles in the World War

When in August, 1914, the sinister black cloud of a world war appeared on the horizon, only the Hun was prepared for the life and death struggle in the air. His formidable fleet of super-Zeppelins had not their match in the world, and his program of airship construction was being pushed forward with the utmost speed and efficiency.

France had the largest fleet of dirigibles among the Allied nations. They were of the semi-rigid type, of only medium size and slow speed. They could not hope to compete on equal terms with the swift and powerful German airships.

Great Britain was far worse off than France, for her airship fleet practically did not exist. The army had only two large modern dirigibles and a few very small vessels like the old *Nulli Secundus*, of little practical value. The navy had no airships at all.

Italy had a few good medium sized vessels, and four large dirigibles were in process of building. Russia, too, had several airships purchased from the other countries, of various makes and types, but she lacked experienced aeronauts with which to operate them.

Both France and England had already made extensive plans for the building of dirigibles, but few of the ships ordered were near to completion in 1914. Only the Prussian was ready for hostilities; his airships gave him a great strategic advantage. By means of them he gained information about the movements of Allied troops and munitions; directed his artillery, bombed Allied positions, and went his way, for the most part unchallenged. His naval airships were likewise a terrible menace. One of them, in the early part of the war, received an iron cross for its work in connection with a German submarine, in an attack on three British cruisers.

Everyone knows of Germany's record in the bombing of cities and

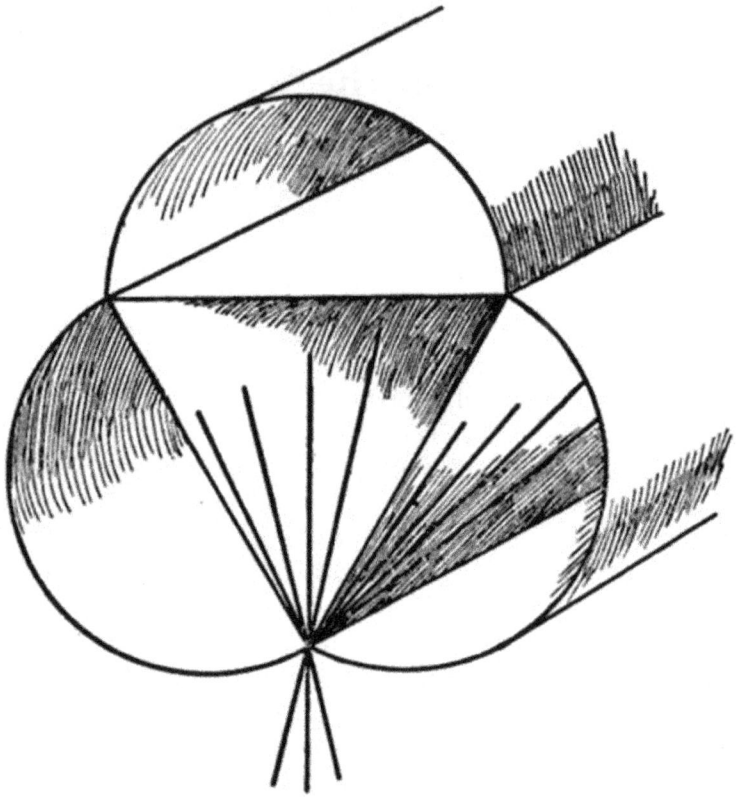

CROSS SECTION OF THE GAS–BAG OF THE *ASTRA-TORRES*, SHOWING METHOD OF CAR SUSPENSION

towns by means of Zeppelins. In the first days of the war the Allies had no anti-aircraft guns and very few airplanes with which to protect themselves, and so Germany went unmolested while she waged her war against defenceless civilians, women and children.

The spirit of the Allies, however, could not be daunted. England put her few small dirigibles on duty over the English Channel, where they served as patrols against submarines. For this work airships are very effective, since it is a curious fact that from their height in the atmosphere it is possible to see far below the surface of the water. So during the first tragic weeks, when France and Belgium were pouring out their life-blood to check the onward sweep of the Hun, these tiny aircraft stood guard over the Channel across which the "contemptible little army" of Britain was being hurried on transports to meet the invader. Like the contemptible little army itself they proved a factor to be reckoned with. Such aerial scouts now form a large arm of the British, French and American navies. Soon after the war began they were constructed in large numbers to serve as patrols against submarines. In the language of the air, these little dirigibles are known as *Blimps*.

The *Blimp* was first developed for use in the war by the British Naval Air Service, but the United States soon saw its advantage as a means of patrolling and guarding our harbours and coastline, and so she set to work to manufacture this type of dirigible in large numbers. Today it is the chief dirigible of our aerial fleet. In some important ways it has the advantage over the airplane in combating the submarine. For the airplane can only remain in the air while it keeps going at high speed. Just as soon as its engines are stopped it commences to descend. But the dirigible can sail out over the harbour, shut off its power and remain motionless in the air for hours, while its observer keeps a constant lookout for enemy undersea craft. When speed is necessary its powerful motor makes it a fast flying craft, sometimes considerably faster than the airplane. For the airplane must often travel against the wind, while the dirigible simply rises until it reaches a current of air moving in the desired direction, when it has the combined power of the wind and its engine to drive it forward.

The U. S. A. *Blimp* is about 160 feet long, rounded in front and tapering to a pointed stern. Its stability and balance are increased by five "fins" at its stern; and it has also four rudders. The car, which is exactly like the ordinary airplane body, has two seats, for pilot and observer, suspended directly from the base of the envelope by wire

THE "BLIMP", C-1, THE LARGEST DIRIGIBLE OF THE AMERICAN NAVY

cables. The *Blimp* carries a 100 horse power Curtiss aviation motor, and is equipped with wireless for exchanging messages.

The French have a small airship very much like the *Blimp* which they use for scout duty. It is called the *Zodiac*, and before the war was designed as a private pleasure car. Because of the fact that it could be easily packed and transported from place to place it was drafted into the service early in the war. Naturally, if an airship has to be kept inflated when not in use it is a constant target for the enemy's gunfire; and a small dirigible which can be packed up in an hour when not needed and readily inflated when the call for action comes is a very much safer proposition.

There are several sizes and slightly different shapes of the *Zodiac*, but the shape of the envelope in all of them is very similar to the *Blimp*, tapering toward the stern with fins to give stability. A large sail-like rudder is set beneath the stern of the ship.

Probably the most interesting thing about the *Zodiac* is the car which in most models has a very long wooden framework. This framework, or girder, by its length distributes the weight along the whole length of the envelope. The car, in which the pilot and observer sit, is set in this girder.

Nothing is more interesting to note in modern airships than the simplification of the method of car suspension. In the early airships the car was hung from the envelope by a large number of cables, which either connected with a network that fitted over the envelope, or else, in a semi-rigid dirigible, to the platform or keel at the base of the balloon.

Now of course all these cables offered a great resistance to the air and were an enemy to speed. Just as the question of speed affected the shape of the envelope, until today we have the streamline balloon, tapering to the rear, and just as it made the question of a rigid or non-rigid envelope so important, it likewise finally did away with complicated connections between the envelope and the car.

From the point of view of car suspension one of the most interesting of the modern French airships is the *Astra-Torres*. This is a dirigible of the non-rigid type. Canvas partitions are stretched across the interior of the envelope in such a manner as to form a triangle, its apex facing downwards. The sides of this triangle are strengthened by cables and from its apex hang the cables which support the car. The air resistance produced by the cables is therefore very slight, since only two lines are exposed.

THE BALLOON OF THE U. S. S. OKLAHOMA

Among the aerial war fleets of the Allied nations, the French offers by far the greatest field for study, since it possesses many different types of dirigibles. The *Astra* and the *Astra-Torres* are perhaps the chief representatives of the non-rigid design, and are generally considered the most successful of the French airships. The *Astra* is the older model, and, like the *Zodiac*, has the long wooden framework or car girder, hung directly to the base of the envelope and distributing to all parts of it the weight of the car. It can be recognized by this and by its stabilizers or small inflated gas bags around the stern of the envelope. The *Astra* is of medium size, varying in length from 199 to 275 feet. The *Astra-Torres* is very much longer, those of the 1914 type measuring 457 feet from nose to stern. From the exterior, this airship has a peculiar three-lobed appearance. It tapers very slightly to the stern and is pointed at both ends, but it has not the *Astra's* inflated stabilizers.

Another French airship of non-rigid design is the *Clement-Bayard*. It is similar in design and in size to the *Astra*, but without the inflated stabilizers. Rounded slightly at the nose, the envelope tapers to a sharp-pointed stern.

The *Lebaudy* is the chief example of a French semi-rigid airship. The envelope is long and cylindrical, pointed at the nose and rounded at the stern, where it is fitted with stabilizing "fins." The base of the envelope is fitted to a long keel, which ends at the rear in a rudder and fins. From this keel the car is suspended by strong cables, and beneath the car extends a conical structure of steel tubes, with points falling downward. These serve as a protection in case of a sudden landing. In front of the car and on each side of the keel are planes similar to those of an airplane, which help to give balance to the ship.

Among airships of the Allies, the French *Speiss* furnishes an example of the purely rigid design. Constructed on the plan of the German Zeppelin, its envelope has an inner wooden framework which holds it in place. The *Speiss* is a large dirigible, measuring about 450 feet. It carries two cars, and in each is a two-hundred horse power motor, giving it great speed.

PART 3

CHAPTER 1

Early Experiments with Heavier-than-Air Machines

For many centuries before the ascension of the first Montgolfier balloon, which, as we have seen, marked the beginning of aerial flight, men had dreamed of a different method of conquering the skies,— in fact, the very natural one suggested by the flight of birds. To build artificial wings was the ambition of many an old-time scientist. Yet practicable as the idea seemed, its working out was, as a matter of fact, beset with difficulties. The Montgolfier balloon rose in the air because it was *lighter* than air,—just as a piece of cork rises in water because it weighs less in proportion to its volume than the water. But a man equipped with wings is a fairly heavy object; where is the force that is to lift him and carry him soaring into the sky?

Unfortunately the early experimenters in aeronautics were not men who had had the long training in keen observation nor the groundwork of mechanical knowledge which would have fitted them for their task of devising a flying machine. They were dreamers and philosophers, often with very clever ideas about how man might succeed in flying. But the exact science of mechanics was yet unborn, and it was not until the nineteenth century, with its great advance in this direction, dawned, that the time was ripe for any measure of success. Still, in many old pictures and medieval manuscripts there are curious examples of the ideas of these old philosophers, designs which were never actually tried out, but which show the longing of men, even in those days, for the great adventure of sailing above the clouds.

All these strange theories of the middle ages were hampered by the superstition that there was some "magic" connected with the power of birds to fly. Cameras were unheard of, or it would have been a simple matter to have recorded on paper the actual motions of the bird's wings in order to study their significance. The astounding ease with which these little winged creatures were able to float across the heavens was indeed baffling; it was difficult to determine just how it was accomplished. Anyone who watches the flight of a seagull realizes that here is an accomplished aeronaut, able to balance himself with perfect ease in the atmosphere, to mount upward on flapping wings, or, taking advantage of a rising air current which can support him, to float motionless with wings extended. All this requires an unusual amount of skill, particularly in balancing. Drop a piece of paper and watch how it turns and tumbles at every angle before it reaches the floor. That is just what a bird or an airplane has a tendency to do, and it takes a perfect system of control and a skilled pilot indeed, to keep it right side up.

The first idea, of course, for a heavier-than-air machine, was that of a pair of wings to be *attached directly to the human body*, and to be worked with the arms. As early as 1480 Leonardi da Vinci drew up a design for an apparatus of this sort. And the idea was not a bad one: it would have worked all very well had it not been for one small fact which the philosophers overlooked, that man is not provided with the powerful shoulder muscles such as the bird possesses for moving his wings.

Altogether, it was not until the nineteenth century that any real progress toward flight in a heavier-than-air machine was made. It came when experimenters began to investigate the definite laws of air resistance and air pressure which control the action of a bird just as they do the action of a kite. As a matter of fact, a bird, or an airplane, is nothing more than a complicated kite, controlled by an intelligence within itself, rather than by an operator standing on the ground and guiding it by means of a cord. Everyone knows that a kite, if placed at an angle to the wind, will be carried upward. The reason for this can be seen from a very simple diagram.

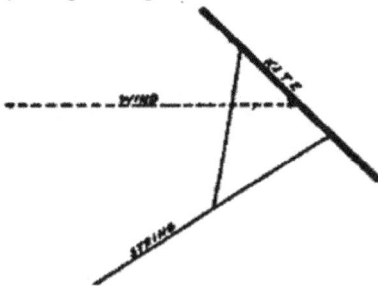

The pressure of the wind would, if unhindered, push the kite into a horizontal position. But the string prevents the angle of the kite from altering, and since the pressure on its lower surface is greater than that on its upper, it naturally rises. This is just what happens when the bird sets his wings at such an angle to the wind that he is lifted into the sky. It is also the principle which governs the airplane or glider, whose planes are kept at a definite angle to the air current. The bird can of course readjust the angle of his wings when he has risen high enough, or when he meets a current of air moving in a different direction, and in the same way the elevating plane of a modern airplane can be lifted or deflected at the will of the flyer, to produce an upward or a downward motion.

The first man to study seriously the effects of air pressure on plane surfaces was an Englishman named Sir George Cayley, who in 1810 drew up plans for a flying machine somewhat resembling the modern monoplane. In 1866 Wenham patented a machine which involved an ingenious idea, that of several parallel planes ranged above each other, instead of the single surface, as of the bird's wing. Wenham believed that the upward pressure of the wind, acting on all these surfaces would give a far greater lifting power, as well as a greatly increased stability, for the machine could not be so easily overturned. Here was the principle of the modern biplane and triplane in its infancy. Yet the idea of strict "bird-form" was more appealing to the imagination, and the experimenters who came after Wenham did not adopt his suggestions.

The man who may truly be said to have given the airplane its first real start in life, was a German named Otto Lilienthal. His figure is a very picturesque one in the long story of the conquest of the air. Lilienthal was a very busy engineer, but from boyhood he had had a consuming interest in the problems of flight, and as he travelled about Germany on his business undertakings he cast about in his mind incessantly for some plan of wings which would support the human body and carry it up into the air. He finally began a very systematic study of the wings of birds with the result that he made some unusual and important discoveries. While the men who had preceded him had attempted only flat wings in their plans for flying machines, Lilienthal decided that the wings should be arched, like those of a bird, heavier in front, with an abrupt downward dip to the front edge, and then sloping away gradually to the rear where their weight was comparatively slight. When still quite a young man he began building kites with planes curved in this manner. To his surprise and joy he found that they rose very rapidly when set to the breeze. They even seemed to move forward slightly in the air, as though they had a tendency to fly. Like a bird resting on a current of air with wings motionless, these little toy wings were carried along gracefully on the breeze. Lilienthal was jubilant. A man equipped with wings like these, he said to himself, would have no difficulty at all in flying.

Lilienthal was not a rich man and it was many years before his opportunity to test his ideas with a real flying machine came. When by hard toil at his profession he had accumulated a comfortable fortune, he turned at last to his beloved study. He had often watched the baby birds in their efforts to fly, and he knew it would be a long time before he attained any skill with wings, but he was absolutely confident that with much practice and perseverance he could actually learn to fly like the birds. So he constructed for himself a pair of bird wings, arched exactly like those which he had studied. They were arranged with a circular strip of wood between them for his body. Here he hung, with his arms outstretched on each side, so that he could operate the wings.

The difficulties Lilienthal had looked for he experienced in large measure. It was no easy thing to attempt to fly in this crude apparatus, but day after day he went out upon the road, turned to face the breeze as he had seen the baby birds do, ran swiftly a short distance, and then inclined the wings upward so that they might catch the current of air. For a long time he was unsuccessful, but imagine his joy when he actually did one day feel himself lifted off his feet, carried forward a few feet and set down. It was scarcely more than a tiny jump, but Lilienthal knew he had commenced to fly. From that time on his efforts were ceaseless. He succeeded in being lifted a number of feet off the ground and carried for some distance. But try as he would he could not get high in the air. He realized that what he lacked was any form of motive power, and for want of a better, determined to make use of the force of gravity to start him through the air at greater speed.

Accordingly he had built for him a hill with a smooth incline, and from the top of this he jumped in his flying machine. The wings he had first constructed he had since improved on, adding two tail planes at the rear which gave greater stability and decreased the tendency to turn over in the air. As he sprang from the hilltop in this curious apparatus, he turned the wings upward slightly to catch the breeze, which supported him exactly as if he had been a kite while he glided out gracefully and finally came gently to earth. This spectacle of a man gliding through the air attracted large crowds. People assembled from far and wide to behold the flying man, and his achievements were greeted with wild cheering. On his huge winged glider he floated calmly over the heads of the astounded multitude, often landing far behind them in the fields. In the difficult matter of balancing himself in mid-air he became exceedingly skilful. Every slight gust of wind had a tendency to overturn him, but Lilienthal constantly shifted the weight of his body in such a manner as to balance himself.

As he gained confidence he began practicing in stronger winds. His great longing was to soar like a bird up into the sky, and so when he felt a rising air current, he inclined his wings slightly upward to take advantage of it. Often he did rise far above the hilltop from which he had sprung, but he never succeeded in actually flying like a bird. His glider had not the motive power to drive it against the breeze with sufficient velocity to send it up into the air, and his wings were but crude imitations of the wonderful mechanism on which the bird soars into the sky. Undaunted by his failure he set to work on a double set of wings, very similar to a modern biplane. He thought these would have greater lifting power, but when he came to try them he found them exceedingly unwieldy and hard to control. For where the bi-plane has an intricate control system, Lilienthal relied entirely upon his own body to operate his glider.

Lilienthal became more and more reckless in his gliding efforts, and in 1896, while gliding in a strong wind, he lost control of his winged contrivance and came crashing to the earth from a great height. When the horrified spectators rushed to the spot, they found the fearless pioneer flier dead beneath the wreck of his machine.

What Lilienthal had done for the cause of aviation, however, would be hard to estimate. He had drawn the attention of thinking people the world over to his experiments. He had pointed the way to the real solution of the problem of flying: that of studying and imitating the birds; and he had discovered the form of plane which on airplanes to-day is well known to give the greatest lifting power: that of an arched surface, deeply curved in front and sloping gradually back to its rear edge where its thickness is very slight. Moreover, his attempts at flight had presented a challenge to engineers and scientists—a challenge which was quickly to bear fruit.

An Englishman named Percy S. Pilcher had followed the work of Lilienthal with the deepest interest, and he now determined to begin a series of experiments on his own account. Like Lilienthal he realized that it would be useless to attempt a motor driven airplane until the principles of glider construction were fully understood. A glider is simply an airplane without an engine, and Lilienthal succeeded in giving it a certain motive power by starting from a high point, so that the force of gravity could draw him forward and downward. Pilcher adopted an even more original scheme for making his glider "go." He treated it exactly as if it had been a huge kite, fastening a rope to it and having it pulled swiftly by a team of horses, until it had gained sufficient momentum to carry it up in the air. The moment it began to rise, Pilcher, who hung between the two large wings much as Lilienthal had done, detached himself from the rope and went soaring into the air like a kite, attempting to balance himself and prevent his glider from overturning. But he had not the experience that long and careful practice had given to Lilienthal, and before he had made very many flights in his glider, he fell and met his death.

In 1896 an Australian, Hargrave, experimented with kites in order to discover a glider form which possessed both lifting power and stability. He was the originator of the familiar "box-kite," which flies so steadily even in a strong breeze. Hargrave connected four very large kites of this sort by a cable, swung a rope seat beneath them and succeeded in making ascents without fear of accident.

Chanute, a Frenchman, now devised a biplane glider with which he succeeded in making brief flights of a few seconds.

The way was now paved for the coming of two great pioneers in the history of aviation. Wilbur and Orville Wright were owners of a small bicycle shop in Dayton, Ohio. They were men with an innate mechanical skill and with the same dogged persistence and indifference to physical hardships which might have brought success to Lilienthal if he had had the time to devote to his experiments.

The Wright brothers had read with fascination accounts of the gliding efforts of Lilienthal. They determined to set to work to solve the problem of human flight. For two years they read and studied everything that had been written upon the subject, and then finally they felt ready to make a trial of a glider of their own construction. They had made up their minds that Chanute's idea of the biplane was most practicable, and so the machine which they built was not strictly bird form, but had two long planes extending horizontally and parallel to each other, attached by wooden supports. The operator or flier lay face downward in the centre of the lower plane.

Their glider was too large to be operated with the arms as Lilienthal's had been, and so they had to devise some new method for controlling and balancing it in the air. This they managed by the use of small auxiliary planes, which were operated by levers and ropes. In front of the two large planes was a small horizontal plane which could be raised or lowered. When raised to catch the wind it gave the glider an upward motion which carried it into the air, bringing the large planes to an angle with the wind where they could continue the climbing process.

One of the great difficulties of the early gliders was their tendency to turn over sidewise. Lilienthal counteracted this whenever he felt one side of his glider falling by shifting his weight toward the highest wing and thus pulling it down. This crude method was impossible in the Wright biplane. The brothers set themselves to seeking a solution from the balancing methods of birds, and right here they made a discovery which was of the greatest importance to the progress of the airplane. The bird when he feels one of his wings falling below the level of the other, simply droops the rear portion of the wing which is lowest, forming a cup or curve at the back which catches the air as it rushes under. This increased pressure of air forces the wing up again until in a second the bird has regained his balance. Imitating this method, the Wright brothers constructed the planes of their glider in such a manner that a cord fastened to the rear sections of each plane could be pulled to draw the rear edge downward.

If the left side of their machine became lower than the right it was a simple matter to pull down the left halves of the rear edges of the two planes, and so catch the air currents which would force that side upward. This ingenious scheme of obtaining sidewise or "lateral" balance is used in a modified form in airplanes today, and is known as "wing-warping."

The brothers chose the coast of North Carolina as the best place for their first attempts to fly, for there the breezes were usually not too strong. After a good deal of difficulty they learned not only to glide, as Lilienthal had done, but also to soar some distance into the air. They had so far no means of turning around, but this was remedied by fastening at the rear of the two large planes a small vertical plane which could be moved from side to side and which served to turn the glider.

There were three achievements in airplane construction which so far could be placed to the credit of the Wrights. One was the *elevat-*

ing plane by means of which an upward or downward motion of the glider was obtained. The second was the ingenious *wing-warping device*, for securing stability. The third was the *rudder*, which enabled the pilot to turn around in mid-air.

Not satisfied with what they had already accomplished, the brothers now turned their attention to constructing a motor suitable for use in a flying machine. This had to be exceedingly light and at the same time strong, and some means had to be discovered for converting its power into motion. The first engine they built was a four-cylinder petrol, and it was used to revolve two wooden propellers acting in opposite directions. The blades of these propellers as they churned the air, gave "thrust" to the airplane exactly as the propellers of a ship drive it through the water. In this new model airplane the flier no longer lay face downward as in the old glider, but sat on a bench between the planes, from which he controlled the action of the engine, the elevating plane, the rudder and the wing warping arrangement by means of levers and cords.

It was in the memorable year of 1903 that this first real airplane was flown by the Wrights. They continued to work steadily upon the problems of design and construction, and after many trials in the next two years, they succeeded by 1905 in building an airplane which would actually fly a number of miles.

They determined to offer their precious secret to some government, and decided on France, which has always been the patron of aviation. But the French government, after an investigation did not accept their offer, and so, disappointed, but still dogged, they retired into silence for a period of several years. In 1908, when their inventions had been patented in every country, they began a series of public demonstrations of their remarkable machine, Orville in America and Wilbur in France.

By that time, unfortunately, other pioneers had stepped forward to claim honours in the field which they first had explored, but the Wright biplane easily outstripped its contemporaries. Their wonderful demonstration flights made them heroes, acclaimed by millions, and their achievements aroused immediate and intense interest in aeronautics.

CHAPTER 2

First Principles of an Airplane

It is almost humorous that man, who for centuries had nourished the secret ambition of acquiring wings, should have found his dream imperfectly realized in the twentieth century by riding in a kite. For that is all an airplane actually is. Yet a "kite" which is no longer tied to earth by a cord and which is equipped with a motor to drive it forward at a great speed has one decided advantage over the old-fashioned sort. The paper kite had to wait for a favorable breeze to catch it up and bear it aloft. We saw in the last chapter how the push of the air against the underneath side of the kite caused it to rise. If instead of the air current pushing against the kite, the kite had pushed against the air, exactly the same result would have been attained. A bird, flying in a dead calm, creates an upward pressure of air by his motion which is sufficient to support his weight. But the bird, as he flies forward against the air creates more resistance under the front portion of his body than under the rear, and this increased upward pressure would be sufficient to turn him over backward if his weight were not distributed more toward the front of his body, in order to counterbalance it.

This fact can be easily illustrated with a piece of cardboard. Take a small oblong sheet of cardboard and mark a dot at its centre. If the cardboard is of even thickness this dot will be the *centre of its weight*. Now hold the cardboard very carefully in a horizontal position and allow it to drop. It should fall without turning over, for it is pressing down evenly on the air at all points. You might say it is creating an upward air pressure beneath it, which is evenly distributed. The *centre* of the supporting air pressure exactly coincides with the centre of weight. If you have not held the cardboard in a precisely horizontal position this will not be true. The unequal air pressure will cause it to lose its balance and "upset." This is very much the sort of experiment that Lilienthal tried when he jumped from the top of a hill in his glider, and it is easy to imagine how much skill he must have required in balancing himself in order to prevent his crude contrivance from overturning.

But now suppose that instead of dropping the piece of cardboard straight down, we give it a *forward push* into the air. As the cardboard moves *forward* it naturally creates more air resistance under the front than under the rear, and this unequal pressure will cause it to do a series of somersaults, before it reaches the floor. The same thing would happen to the bird or the airplane whose weight was evenly and equally distributed.

Now since the air pressure is greater under the front of the cardboard, add a counterbalancing weight by dropping a little sealing wax at the centre front. The dot that you made in the middle of the sheet is no longer its centre of weight. The *centre of weight* has moved forward, and if it now corresponds to the *centre of pressure* the cardboard can be made to fly out and across the room without overturning.

The whole problem of balancing a glider or an airplane is simply this one of making the centre of weight coincide with the centre of the supporting air pressure. Adding weight at the front of the glider is not the only way of doing this: perhaps the reader has already thought of another. Since the air pressure is caused by the weight of the cardboard and its forward motion, we could cut the sheet smaller at the front so as to lessen its air resistance there, or we could add a "tail" at the stern in order to create more air resistance at that end. Either of these plans would move the *centre of pressure* back until it corresponded with the *centre of weight*, and so would complete the balance of our cardboard glider.

In the bird's body all of these methods of obtaining balance are combined. His body and head taper to a point at the front in order to decrease the forward air resistance. The weight of his body is distributed more toward the front, thus counterbalancing any tendency to whirl over backward. His tail increases the stern resistance, thus helping to draw the centre of pressure back to correspond to the centre of weight.

We begin to see some reasons why a man equipped with wings could never be taught to fly,—as well as how perfectly the form of the bird is planned to correspond to his mode of travel. No wonder the early experimenters with wings, finding themselves so utterly helpless and awkward, attributed the bird's ease and grace of carriage to

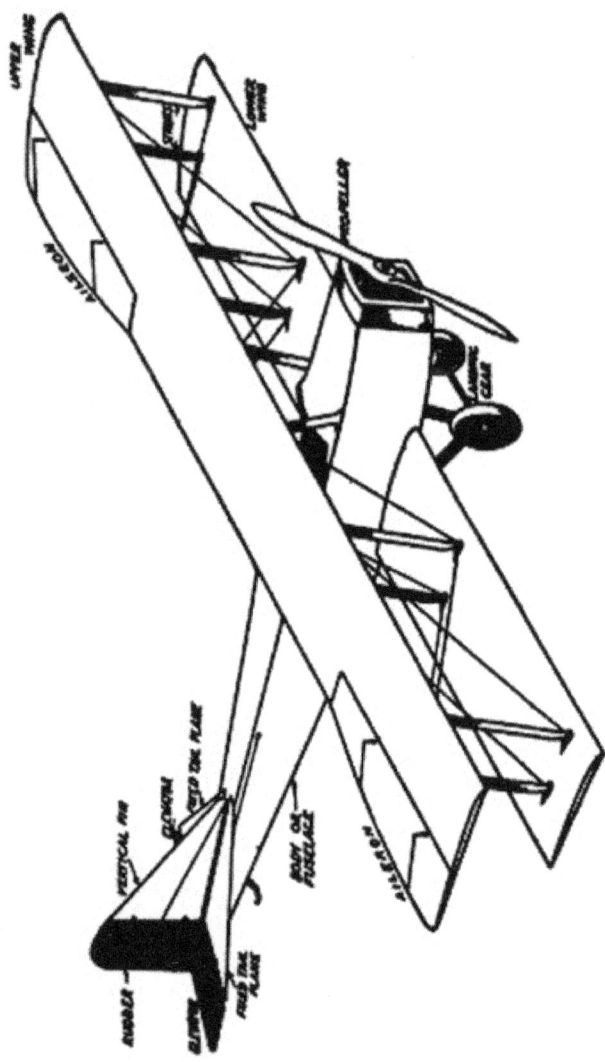

UPPER WING

STRUT

LOWER WING

PROPELLER

AILERON

LANDING GEAR

VERTICAL FIN

ELEVON

FIXED TAIL PLANE

BODY OR FUSELAGE

AILERON

RUDDER

ELEVON

FIXED TAIL PLANE

DIAGRAM SHOWING THE ESSENTIAL PARTS OF AN AIRPLANE

"magic."

The modern airplane is constructed with the most painstaking attention to this principle of *balance*. Next to it in importance is that of *wing construction*: that is, the size, shape and proper curve of the supporting planes. Here again the construction of the airplane follows very closely the general form of the bird. A large bird which flew very high would be found to have his wings arched high in front, where they would have considerable thickness, and sloping down very rapidly toward the rear, while their thickness rapidly diminished. This sort of wing has great lifting power, and it is the sort that is used on an airplane which is built to "climb" rather than to develop speed.

As the arched wing cuts through the air it leaves above it a partial vacuum. Nature always tends to fill a vacuum, and so the airplane is drawn upward to fill this space. As the wings cut through the air a new vacuum is constantly created and so the airplane mounts higher and higher. The airplane is being carried upward by two forces: the air pressure beneath it and the vacuum above it which draws it up. The air pressure beneath it increases with the speed at which the airplane is travelling, and it has a tendency to press the wing into a more horizontal position, thus destroying its climbing properties. At the same time, when this happens, the thick front section of the wing presents a great "head resistance" which retards progress, and a very high speed becomes impossible.

Wings of this type can never be used on an airplane which is intended to travel at high speed. They were used on the heavy bombing and battle planes of the Great War, for they are capable of lifting a very great weight. But on the scouting planes, where speed is essential, a totally different sort of surface was employed. Here the plane is very little arched and of almost even thickness, tapering only very slightly to the rear edge. It also tapers somewhat at the front, so as to lessen its "head resistance" as it cuts through the air.

Such a surface creates little vacuum above it, and consequently has not a great lifting power. On the other hand it offers little "head resistance" and so permits a high speed. And right here it should be mentioned that a powerful motor does not in itself make a swift airplane, unless the wings are right,—for if the wings create a strong resistance *in front of the airplane* they destroy speed as fast as the motor generates it.

Remember that the lifting power of the airplane wing is made up of two factors. *First*, there is the resistance or the supporting air pressure created by the weight and speed of the wing; *second*, the arch of the wing creates a vacuum above it which tends to lift the airplane up. Now when for speed the arch is made very slight, the lifting power can still be increased by increasing the *area* of the wing, thus adding to the upward pressure. Thus for certain war duties an airplane with very large, comparatively flat wings can develop both a very good lifting power and a very high speed.

We have already mentioned the "head resistance" of the airplane wing. If the wing could strike the air in such a way as to sharply divide it into currents flowing above and below, there would be no head resistance. But the very arch of the wing in front gives it a certain amount of thickness where it strikes the air, so that instead of flowing above or below, a portion of the air is pushed along in front, retarding the progress of the airplane. This resistance is called by aviators the "drift." The best wing is the one which has the maximum lifting power with the minimum head resistance, or, to use technical language, the greatest "lift" in proportion to its "drift."

Of course, not only the wing but all parts of the airplane offer resistance to the air. In order to reduce this total head resistance to the minimum, every effort is made to give the body or "fuselage" of the airplane a "streamline" form,—that is, a shape, such as that of a fish or a bird, which allows the air to separate and flow past it with little disturbance. For this purpose the fuselage of the airplane is usually somewhat rounded and tapering toward the ends, often "egg shaped" at the nose.

The method of "wing warping" invented by the Wright brothers is still used on all modern airplanes to preserve lateral stability. The part of the wing which can be warped is called the *aileron*. There are two ailerons on every wing, one on each side at the rear, and they may be raised or drawn down by the action of a lever operated by the pilot.

If the pilot feels that the left side of his machine is falling, he draws down the aileron on that side and raises the right hand aileron. The aileron which is lowered catches the air currents flowing beneath the wing on that side. At the same time the raised aileron on the right lessens the pressure under the wing on that side and so gives it a tendency to fall. In this way, in a fraction of a minute the wings are brought level again and lateral stability is restored.

Whereas the old Wright biplane had an elevating plane in front of the main planes, most machines today have the elevating surfaces at the rear. By raising the "elevators" an upward motion is obtained, or by lowering them, a downward motion.

Steering to right and left is accomplished by a rudder at the rear of the airplane body or "fuselage." This rudder may be turned to right or to left, working on a hinge.

WRIGHT STARTING WITH PASSENGER

AN EARLY FARMAN MACHINE PRIOR TO START

CHAPTER 3

The Pioneers

While the Wright brothers, lacking both funds and encouragement to continue their remarkable project, remained, from 1905 to 1908 in almost total obscurity—their wonderful flying machine packed away ignominiously in a barn,—in France a number of eager experimenters were working assiduously to outstrip them, and it was only by great good fortune that when Wilbur Wright arrived in France in 1908 he did not find himself beaten from the field. Actually the Wright machine was far in advance of the early French models, and although the French, with true spirit of sportsmanship, were quick to admit it when the fact was demonstrated, yet prior to 1908 they had no idea that such was the case, and were enthusiastically proud of their home-made models.

Among the very first of the French pioneers of flight was that gallant little Brazilian, Santos-Dumont, whose exploits with the dirigible had done so much to popularize air sports. His name was a household word with the French, who literally lionized him. Impatient of the limited opportunities for adventure presented by the dirigible, Santos-Dumont cast about in his mind for some means of procuring a more agile steed on which to perform his aerial tricks. In 1904 he became deeply interested in the subject of gliding, and made up his mind to try a few gliding experiments of his own. Like everything else he had attempted his method of attacking this new problem was startlingly original. Lilienthal and the other gliders had all made their flights above the solid ground. Santos-Dumont liked the idea of rising from the water much better.

He ordered built for him a glider of his own design for this particular purpose. On every clear day when the wind was favourable, the plucky little aeronaut was out, learning to use his new-found wings.

His glider, which floated on the surface of the water, had to be towed swiftly for some distance by a boat in order to give it the initial speed which Lilienthal secured by taking advantage of the force of gravity in his downward jump from the hilltop. Once he felt his speed to be sufficient, Santos-Dumont gently inclined his wings upward to catch the air current. To the surprise of everyone he was remarkably successful. He actually succeeded in soaring short distances, and after a series of efforts he acquired a fair amount of skill in the use of his glider apparatus.

The next step was to attach some motive power to his flying machine. Before very long he had ready for trial a much more pretentious biplane glider, equipped with an 8 cylinder motor which drove a two-bladed aluminum propeller, and fitted with several original appliances to increase its soaring powers and its stability. In front was a curious arrangement resembling a box-kite, which was intended to fulfil the same purpose as the elevating plane which the Wright brothers placed in front of the two main planes of their machine. Santos-Dumont had experienced the same trouble as all the other gliders: the difficulty of keeping his machine in a horizontal position. The tiniest gust, blowing from one side or the other, was sufficient to cause it to lose its balance, and over it would topple sidewise. To overcome this obstacle the Wright brothers had adopted the ingenious method of wing-warping, imitated directly from the habits of birds. Santos-Dumont was not nearly of so scientific a turn of mind as the two great American pioneers.

Without having gone so deeply into the subject, he determined to place upright planes between his main planes, to ward off gusts and increase the lateral stability. The idea was not a bad one, though far from being the best. In the summer of 1906 he flew with his glider successfully very short distances. In October of the same year he accomplished *a demonstration flight of 200 feet* at Bagatelle, near Paris. At the present day when airplanes go soaring above our heads faster than express trains, making long, continuous cross-country flights, that journey of 200 feet seems humorous, but at the time it was the European record. It aroused a great deal of popular enthusiasm, for the French, with their vivid powers of imagination, were quick to see the possibilities in this new, heavier-than-air contrivance. At once the Brazilian set to work to outstrip this first achievement. This time his originality took an entirely new turn. Instead of the biplane type he decided on a monoplane, and he began laying out plans for a monoplane so tiny, yet so efficient, that it was destined to become famous. But it was several years before this miniature flier was ready, and so for a while the idol of the French public dropped almost completely out of sight.

In the meantime others were up and doing in France. Henry Farman, who already had made his name famous in motor car racing, was the next to win popular acclaim for exploits in the air. Farman was known as a man of the most consummate daring, cool-headedness in emergency, and quick judgment. An Englishman by birth, he had resided all his life in France, where with his brother Maurice he had achieved an enviable reputation as a sportsman. Farman afterward designed and constructed airplanes of his own, but it was in one built by the Voisin brothers that he first took to the air.

The Voisins were very ambitious indeed in their first airplane project. The machine which they built was both large and heavy, and possessed of many unscientific features. Like the Wrights' machine it had two large horizontal planes, in front of which was placed a small elevating plane, which could be inclined up or down to lift the airplane into the air or bring it to earth again. Unlike the Wright model it had a large "tail," or horizontal plane at the rear, intended to give it increased longitudinal stability. This feature represented an improvement. The Wrights had to keep their machine on the level by raising or lowering the front elevating plane in such a way as to counteract any pitching motion, but the tail of the Voisin biplane gave it a great deal more steadiness in the air. Fitted to the tail was a rudder, by which turning to right or left was accomplished. But the Voisin brothers had no wing-warping device on their large flier. Instead they used the upright curtains or planes between the main planes, which we have already seen on the machine designed by Santos-Dumont. Their airplane was equipped with an 8-cylinder motor, which turned a large propeller.

In this large and unwieldy machine, weighing possibly 1400 pounds, Henry Farman made a short flight in a closed circuit in 1908. At the time it was the record flight in Europe, and the French people fondly imagined it was the best in the world. That same year Wilbur Wright arrived on French soil and showed them in a few astounding experiments what the Wright biplane could do.

The successes of this tall, untalkative American, who had come over to France and with ease made the aerial adventures of Santos-Dumont and Farman seem like the first efforts of a baby learning to crawl, greatly as they surprised, and, perhaps, disappointed the French people, in the outcome had the result of spurring Frenchmen on to greater effort in the problem of airship design. Before the end of 1908 Henry Farman, in an improved Voisin, had wrested back the lost honours by flights which were longer than those made by Wilbur Wright.

And other Frenchmen were hard at work. After building a number of machines and meeting with many accidents and failures, Blériot emerged in the summer of 1909 with a successful monoplane. At almost the same time the Antoinette monoplane made its appearance, and soon these two similar machines were pitted against each other in a famous contest.

The London *Daily Mail*, with the intention of stimulating progress in aviation, put up a prize of £1000 for the first machine to fly the British Channel. In July, Blériot brought his monoplane to Calais; and Hubert Latham appeared as his antagonist, with an Antoinette machine. Both of the contestants were skilled pilots, and both were men of fearless daring. The feat which they were about to attempt required men with those qualities, for in these pioneer days of aviation it was not the easy task to fly the Channel which at first glance it might seem to be. Over the Channel the winds were almost always very severe, and they represented the greatest danger the airman had to face. The first airplanes had so small a factor of stability that it was almost impossible to fly them in even the gentlest breeze.

The most intrepid aviators never once thought of attempting flight in unfavourable weather. To be overturned in crossing the Channel meant taking a big risk of death, and both Blériot and Latham realized that they were taking their lives in their hands in undertaking the trip. They had a long wait for calm weather, but on July 24th conditions seemed right for a start the next morning. Just at dawn Latham flew out across the sea and disappeared in the distance. Not very long behind him, Blériot, having tested with the utmost care every part of his little machine, climbed into the pilot's seat, and with a "Goodbye" to the little group of mechanics and friends who stood about, sped away, hot on the trail.

On and on flew Latham in his larger Antoinette monoplane, and the hope of victory began to loom big. Far out over the Channel however, his engine suddenly "went wrong," as engines in those days had a habit of doing, and the much feared thing happened: he began to fall. In a very few moments the plucky pilot was clinging to his airplane, as it floated for a few moments on the choppy sea. Before it could sink a vessel had hurried to the rescue, and Latham was hauled on board, disappointed, but safe.

Blériot, meanwhile, was far from being sure of his course as he flew on steadily through the early morning haze. But his engine continued to run smoothly, and finally far ahead, the white cliffs of England began to emerge out of the distance. With joy in his heart the Frenchman flew proudly in over the land and brought his airplane to the earth in the vicinity of Dover Castle. He was greeted as a hero by the British and the glad message of his triumph was speeded back to Calais.

Loath to be behindhand in airplane activities, America was also busily at work developing the heavier-than-air machine, and another famous name had by this time been added to that of the Wright brothers. By 1909 Glenn Curtiss with a group of distinguished co-experimenters had succeeded in constructing several very interesting flying machines. Curtiss' story is an interesting one. In 1900 he was the owner of a small bicycle shop in Hammondsport, New York. He had a mania for speed, having ridden in many cycling races, and it was he who first thought of attaching a motor to a bicycle for greater speed. He soon sprang into the limelight as a motorcyclist and a manufacturer of motorcycles. A small factory went up at Hammondsport, and achieved a reputation for the very good motors it turned out.

Curtiss first became interested in flying through an order he received from Captain Thomas Scott Baldwin for a motor to be used in a dirigible balloon. He set to work on the problem of constructing a motor suitable for the purpose, and, as might be expected, he became fascinated with the possibilities of flight. Curtiss and Baldwin made some very interesting experiments with the dirigible. Then, in 1905, Curtiss made the acquaintance of Dr. Alexander Bell. The famous inventor of the telephone was engrossed in the study of gliding machines, and had been carrying on a series of experiments with kites by which he hoped to evolve a scientific airplane. To further these experiments he had called in as associates in the work two engineers, F. W. Baldwin, and J. A. D. McCurdy, while Lt. Thomas Selfridge of the U. S. Army was also greatly interested.

Thus it came about that in the summer of 1907 this group of capable men formed what they were pleased to call the "Aerial Experiment Association," of which Curtiss was perhaps the moving spirit. The first machine built by the Association was christened the *Red Wing*, the second the *White Wing*; the third was called the *June Bug*, and it proved so successful a flier that on July 4th, 1908, it was awarded the *Scientific American* trophy for a flight of one kilometre, or five-eighths of a mile.

While, in France, Farman and the Voisin brothers, Latham and Blériot were pushing steadily along the rough road to aviation successes,—in America, the Wright brothers and Curtiss with his associates, were demonstrating to the public on this side of the water what flying machines could do.

In fact, the airplane had definitely begun to assert its superiority as master of the air, and many eyes in all parts of the world were fixed on it and on the great future possibilities for which it stood. Everywhere, warm interest had been aroused, and, at least in France, the military importance of the heavier-than-air machine was coming to be realized.

Now the time was ripe for the great public demonstration of the world's airplanes which took place at Rheims in August, 1909. The Rheims Meeting is probably the most memorable event in the history of aviation. It placed the work of a dozen or more earnest experimenters definitely in the limelight, and gave the chance for comparisons, for a summing up of knowledge on the subject of flight, and for a test of strength, which resulted in the mighty impetus to aerial progress which followed immediately afterward.

Here at Rheims were gathered many famous flying men who already had made their names known throughout Europe and America. There were Farman, Latham, Paulhan, Blériot, Curtiss, and the three who flew Wright machines, the Comte de Lambert, Lefevre and Tissandier,—as well as many others, for there were thirty contestants in all. Many unusual feats delighted the spectators. Lefevre, a student of the Wrights, and up to that time unknown, amazed the assemblage by his wonderful aerial stunts. He circled gracefully in the air, making sharp, unexpected turns with the utmost skill, and winning round after round of applause.

Curtiss and Blériot emerged as contestants for the speed prize over 10 kilometres, and after several breathless attempts in which records were made and broken, the honour was finally carried off by Blériot, who covered the distance of 10 kilometres (about 6¼ miles) in 7 minutes, 47.80 seconds. Curtiss replied by beating his famous opponent in the contest for the Gordon Bennett Cup, offered for the fastest flight over 20 kilometres; and Curtiss also was the winner of the 30 kilometre race.

It was Farman, in a biplane of his own design, who surprised everyone by his remarkable performance, and turned out to be the victor of the occasion. Flying for three hours without stopping, round the course, he covered 112 miles without the slightest difficulty, and was only forced to make a landing because of the rapidly approaching dusk. For his feat he was awarded the Grand Prize, and was hailed as the most successful of all the contestants.

Finally Latham, in an Antoinette monoplane, proved he had the machine with the greatest climbing powers, and carried off the Altitude prize on the closing day of the meeting.

Among those who looked on at the famous Rheims Meeting of 1909 there were none more keenly and intelligently interested than the representatives of the French military authorities. They had come for two reasons: to ascertain at first hand which were the best machines and to order them for the French Government; on the other hand, to encourage to the fullest extent possible all those men present who were earnestly working in the interests of aviation. France was ready and willing to spend money freely for this purpose, and the Rheims Meeting resulted in orders for machines of several makes. Some of these were regarded as having great possibilities from a military point of view; and others, though not looked on so favorably, were purchased as a sign of goodwill and support to future experiment. It was this far-seeing patronage which paved the way for France's later aerial triumphs, for it gave her a diversity of machines and a devoted coterie of workers all following original lines of experiment.

Let us glance for a moment at the little group of machines which stood out by their merits most prominently at that Rheims Meeting of 1909, and which gave the greatest promise for the future. Today they seem antiquated indeed, but for all their rather curious appearance they were the legitimate forefathers of our powerful modern airplanes. Among the biplanes, those especially worthy of note were the Farman, the Wright, and the Voisin; while the Blériot and Antoinette monoplanes gave a most excellent account of themselves.

Farman, who had first learned to fly in a machine designed and built by the Voisin brothers, was far from satisfied with his sluggish, unmanageable steed and at once set to work on a design of his own. His one idea was to construct a biplane of light weight, speed and general efficiency. He did away with the box-kite tail of the Voisin model and substituted two horizontal tail planes with a vertical rudder fitted between them. Instead of the vertical planes or "curtains" between the main planes by which the Voisins attempted to preserve the lateral stability of their airplane, Farman adopted the "wing-warping" plan of the Wrights in a somewhat modified form. The Wright machine, it will be remembered, had wings whose rear portions were flexible, so that they could be drawn down at the will of the pilot. If the latter felt that the left side of his machine was falling he simply drew down or "warped" the rear edges of the wings on that side. The air rushing under the wing was blocked in its passage and the greater pressure thus created forced the wing upward on the left side until balance had been restored. Acting on this principle, Farman attached to the rear edges of the main planes at each side a flap, or as it is called today, an *aileron*, which worked on a hinge, so that it could be raised or lowered.

Another novel feature of this first Farman biplane was its method of starting and landing. Below the planes had been placed two long wooden skids, and to these small, pneumatic tired wheels had been attached by means of strong rubber bands. In rising, the airplane ran along the ground on these wheels until it had acquired the momentum necessary to lift it into the air. When a descent was made, the force of contact with the ground sent the wheels flying upward on their flexible bands, and allowed the strong skids to absorb the shock. This underbody or *chassis* was a distinct improvement on anything that had yet been devised, for it was light in weight and efficient.

In one other important respect the Farman machine was superior to all those demonstrated at Rheims in 1909, and that was in its engine. Airplane engines up to this time had been nothing more or less than automobile engines built as light in weight as possible. But in France a new engine had made its appearance, designed especially for airplane needs. Hooted as a freak at the first, and rejected by experts as "impossible," it carried Farman round the course on his three hour flight without a hitch and made him the winner of the Grand Prize. This remarkable engine was the Gnome and the reason for its excellence lay in its unusual system of cooling. The overheating of his motor was a thorn in the flesh of many an early aviator.

An engine which gave good service in an automobile would invariably overheat in an airplane because of the constant high speed at which it must run. Now motor car engines of whatever type, and whether water-cooled or air-cooled, had fixed cylinders and a revolving crankshaft. In the Gnome motor the cylinders revolved and the crankshaft was stationary. Flying through the air at tremendous speed they necessarily cooled themselves. This was the secret of the perfect running of the Farman biplane. Though Farman had been the first to recognize the merits of the Gnome and install it in his machine, he was not the last, for after the Rheims Meeting it rapidly became the favourite of practically all builders.

Next to the Farman, the Wright machine was probably the best for all-around service of the many demonstrated at the great meeting. Its one greatest disadvantage was the fact that it had to be launched from a rail. It carried no wheels—merely skids for landing—and so to gain initial momentum it had to be placed on a small trolley which ran down a rail. Such a method of gaining speed was exceedingly complicated, and the question at once arises: What would the pilot do if forced to make a landing far from his starting point? Of course it would have been quite impossible for him to have risen into the air for a return trip, and his machine, though in perfect condition, would have to have been packed and carted back home.

The Voisin biplane, though improved since Farman had piloted it in 1908, was still in 1909 an overly heavy, slow flying machine, more or less difficult to steer. It still had its "box-kite" tail and its upright curtains between the main planes. And it carried a rather weighty landing chassis built of hollow metal tubing, to which were attached pneumatic-tired bicycle wheels. Small wheels were also placed under the tail, to support it when running along the ground.

The Blériot monoplane could have claimed the honours for *simplicity*. It had a body built up of light woodwork, over part of which fabric had been stretched. On either side of the body extended the two supporting planes, supported above and below by wires. In the front of the body was the engine and at the rear extremity a small stabilizing plane. At the ends of the stabilizing plane, on either side, were two small planes which could be moved up and down. They took the place of the front elevating plane employed on the other machines. Just behind the stabilizing plane was the vertical rudder, which turned to right or left. The wings of the Blériot had the Wright brothers' wing warping arrangement. The pilot sat just behind the engine, operating the controls.

Larger in wing span and longer in body than the Blériot was the Antoinette monoplane. Like the Blériot it had its elevating planes at the rear, and carried its engine in the bow. Instead of the wing warping device it made use of movable flaps or *ailerons* at the rear edges of the wings. Another idea had been incorporated in this machine for the purpose of maintaining lateral stability. Its wings, instead of ex-

WRIGHT MACHINE RISING JUST AFTER LEAVING THE RAIL

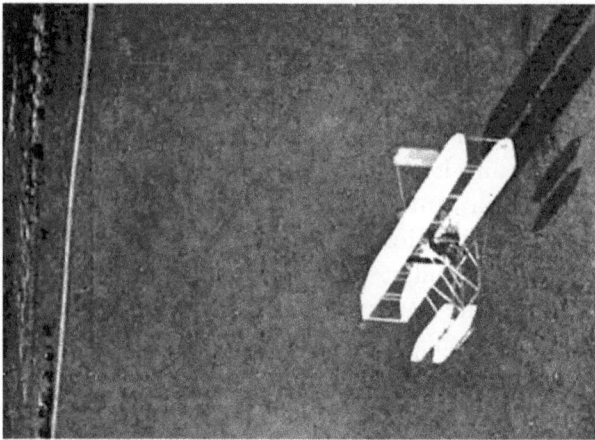

AN EARLY WRIGHT MACHINE, SHOWING ITS METHOD OF STARTING FROM A RAIL

tending in a horizontal position from the body were inclined slightly upward,—a plan which met with serious condemnation from the engineering experts.

These five then, were the machines which claimed most attention in 1909, although many others,—as for instance the R. E. P. monoplane, built by M. Esnault-Pelterie, and the Breguet biplane—were flown at the famous meeting.

The Rheims event had been hugely successful, and the news of the splendid achievements of the airplane spread like wildfire throughout the world. Smaller meetings were arranged for in other cities, and everywhere the great aviators were called for to give exhibition flights. In September Santos-Dumont came once more before the public with the tiniest monoplane in existence, a little machine which he called the *Demoiselle*, and in a series of experiments proved its remarkable capabilities. Santos-Dumont had been residing for some time at St. Cyr, where he had worked on his designs for the *Demoiselle*. One of his aviator friends, M. Guffroy, was also experimenting at Buc, five miles away. The two men agreed that the one who first completed an airplane should fly in it to the home of the other and collect £40. In 6 minutes and 1 second Santos-Dumont covered the five miles on the 14th of September and claimed his reward.

Orville Wright at about this time was exhibiting his airplane in Berlin and winning new laurels before the Crown Prince and Princess of Germany. By the middle of October he was in France, and was present at the Juvisy Meeting, when the Comte de Lambert, leaving the course unexpectedly, made his sensational flight over Paris, circling round the Eiffel Tower at a height of 1,000 feet. Paris was filled with amazement and delight at the sight of an airplane soaring over the city. It was almost an hour before the Comte de Lambert, flying with the greatest ease, arrived once more at the course, to be overwhelmed with congratulations.

On November 3rd, Henry Farman made a world's record of 144 miles in 4 hours, 17 minutes and 53 seconds, wresting from Wilbur Wright the coveted Michelin Cup. In December Blériot attempted an exhibition of his monoplane in Constantinople, but his machine lost its balance in the severe wind which was blowing and came crashing to earth. Though severely wounded, the great aviator recovered rapidly, justifying the oft-repeated superstition that he was possessed of a charmed life.

Thus the year which had meant so much in the forward march of

THE PROPELLER DEPARTMENT IN ONE OF THE GREAT CURTISS FACTORIES

aviation drew to a close. Beginning at Rheims, the reputation of the heavier-than-air machine had spread in ever widening circles throughout all civilized lands. Most important of all, the military authorities of several nations had opened their eyes to tremendous importance of the airplane as an implement of warfare, and their realization of this fact was destined to bring about new and weighty developments within the next few years. Among the great European states only one nation slept while the rest were up and doing, and she saw the day when, with the shadow of war looming on the horizon, she had cause for bitter regrets.

The beginning of 1910 saw the famous aviator Paulhan in the United States for a series of exhibition flights. On January 12th he made a world's record for altitude, climbing at Los Angeles to a height of 4,140 feet, in a Farman machine.

In the Spring there occurred in England a memorable contest between Paulhan and a young flier who up to that time was unheard of, but who rapidly made a reputation for himself in aviation. The London *Daily Mail*, which had already done so much to arouse enthusiasm for the airplane in the British Isles, now offered a prize of £10,000 for the first cross-country flight from London to Manchester. There arose as England's champion Claude Grahame-White, and Paulhan with his Farman biplane was on hand to dispute the honours with him. The distance to be covered was about 183 miles, and the task seemed almost impossible, largely owing to the nature of the country over which the flight must be made. It was rough and hilly and thickly sprinkled with towns, making the task of a forced landing a very perilous one. Engines in 1910 were none too reliable and were apt to play strange tricks. To be forced to descend over a town or in rough country meant a chance of serious accident or death. Rough country moreover is apt to be windy country, with sharp, unlooked-for gusts blowing from unexpected quarters. It was these above all things which filled the airman's heart with dread, for he knew only too well the limited stability of his pioneer craft.

Late in the afternoon of April 27th, Paulhan, whose biplane, in perfect repair, was awaiting him at Hendon, near London, ascertained that the wind was favourable, and at once rose into the air and started on his long trip. Grahame-White had assumed that it was too late in the day to make a start, and had left his machine, all ready for flight, at Wormwood Scrubbs, intending to make a start in the early morning. Shortly after six the news was brought to White that Paulhan was on his way, and he immediately rushed to his starting point and hurried after his rival.

Paulhan had studied every inch of the ground and knew what conditions to expect. His earlier start gave him a great advantage, for he managed to get farther before nightfall, and also before any adverse winds arose. With darkness both pilots were forced to make landings, but Paulhan was far ahead, and the prospect of victory began to wane for the plucky young English flier. In the emergency he determined on a desperate attempt to overcome his handicap. Night flying then was a thing unheard of, but Grahame-White prepared to try it, however risky. At half past two in the morning, by the wan light of the moon he arose from the field where his machine had been landed and flew off into the murky night.

Disappointment awaited the dauntless pilot, however. He had a stern struggle with the wind, his engine began to give trouble, and finally he was compelled to come to earth.

Paulhan got away at dawn and being the more experienced pilot of the two, managed, after a sharp tussle with the wind, to arrive intact at his destination. He was greeted with wild enthusiasm and was indeed the hero of the day.

But England was not without gratitude to her defeated airman, who in the face of enormous difficulties, had persisted so gallantly in his effort to uphold his country's honor in the records of aviation. Though official England was slow to recognize the airplane's claims, the British public showed keenest interest in all the exploits of their sportsmen of the air, and before long there was quite a fair-sized group of such men demanding attention.

America also had a remarkable feat to record in the summer of 1910. The New York *World* had offered a $10,000 prize for a flight down the Hudson River from Albany to New York. The difficulties were even greater than those of the London-Manchester contest, for here the airman had to fly the entire distance over a swift stream. The high hills on either side meant increased peril, for there were sure to be powerful wind gusts rushing out between the gaps in the hills and seeking to overturn the machine. If the engine should give out, there was no place to land except in the water itself, with slight chance of escape for either the pilot or his airplane.

Nevertheless, Glenn Curtiss, whose accomplishments at the Rheims Meeting we have already witnessed, determined to try for the prize. His machine was brought from Hammondsport to Albany ready for a start, and on May 31, after a long wait for favourable atmospheric conditions, he was on his way. A special train steamed after him, carrying newspaper reporters and anxious friends, but he left it far in the distance while he flew swiftly down the Hudson. Villagers and boatmen waved and shouted to him as he passed. At one point he encountered an air "whirlpool" that almost sucked him down, but he succeeded in righting his machine and getting on his way again. Near Poughkeepsie he made a landing to obtain more fuel, and from there he flew straight on to his journey's end, reaching New York City and descending in a little field near Inwood.

In July of 1910 came the second big Rheims Meeting, to show what unprecedented advances had been made in one short year. Almost 80 contestants appeared, as compared with the 30 of 1909. Machines were in every way better and some very excellent records were made. The Antoinette monoplane flew the greatest distance (212 miles), and also reached the greatest height; while a new machine, the Morane monoplane, took the prizes for speed.

Meanwhile the French Army had been busy training aviators and securing new machines. In the Fall these were tried out at the Army Maneuvers in Picardy, and for the first time the world saw what military airplanes really could accomplish. In the sham warfare the army pilots flew over the enemy's lines and brought back astonishingly complete reports of the movements of troops, disposition of forces, etc. The French military authorities themselves, enthusiastic as they had been over the development of the airplane, had not anticipated such complete success. They were delighted with the results of their efforts, and a strong aerial policy was thereupon mapped out for France.

England at this date possessed *one* military airplane, and it was late before she awakened to the importance of aviation as a branch of warfare.

Germany, Italy, Russia, and America were looking on with keen interest, but for a while France maintained supremacy over all in her aerial projects. By the end of the following year she had over 200 military machines, with a competent staff of pilots and observers.

To follow the course of aviation achievement we must now go back to England, where in July, 1911, another big *Daily Mail* contest took place. This time the newspaper had put up a prize of £10,000 to be won by flying what was known as the "Circuit of Britain." This had been marked out to pass through many of the large cities of England, Scotland and Ireland. There were seventeen entrants for the contest, which was won by a lieutenant of the French Navy, named Conneau. Cross-country flights were growing longer and longer, keeping pace with the rapid strides in the development of the airplane. Still another contest during 1911 was the "Circuit of Europe," which lay through France Belgium and England; while a flight from Paris to Rome and one from Paris to Madrid served to demonstrate the growing reliability of the aircraft.

Money had always flowed freely from French coffers for this favourite of all hobbies. At the Rheims Meeting in October of 1911 the government offered approximately a quarter of a million dollars in prizes for aerial feats and in orders for machines. Representatives from many countries visited the meeting to witness the tests of war airplanes.

In the two years since the first Rheims Meeting many vast changes had taken place. Pilots no longer feared to fly in high winds; machines were reliable, strong and swift. A number made non-stop flights of close on to 200 miles, and showed as well remarkable climbing abilities.

It was the Nieuport monoplane which led all others at this Rheims Meeting. Today the name of Nieuport is familiar to everyone, for the little scout machines carried some of the bravest pilots of France and America to victory in the air battles of the Great War. Even in 1911 the Nieuport monoplane was breaking all records for speed. Carrying both a pilot and a passenger it flew as fast as 70 miles an hour at Rheims.

Another new machine that attracted attention was the Breguet biplane, a heavy general service machine weighing 2420 pounds and carrying a 140 h. p. Gnome motor. The Gnome had so far outdistanced all competitors that it had virtually become the universal motor for airplanes, and, many of those seen in 1911 were equipped with it. Since then vast improvements have been made in stationary engines but at that time they almost entirely failed to meet the requirements of light weight, high power and reliability.

One development in the biplanes of 1911 cannot be passed over, for it bears a very interesting relation to their efficiency as war machines. Anyone who has seen a photograph of one of the early biplanes must have been struck by the curious kite-like appearance it presented, due to the fact that it had no *body* or fuselage, but only two large planes, connected by strong wooden supports, and usually with a seat for the pilot in the centre of the lower plane.

It was in the monoplane that a car or airplane body first made its appearance, and to it the wing surfaces of the monoplane were strongly braced with wires. Many of the biplanes of 1911 had adopted the idea and in consequence began to take on a more modern appearance. It was a thoroughly good idea, for by means of its greater stability and strength, protection for the pilot and general efficiency were obtained. Biplanes of this type now carried their engines in the fuselage bow with the pilot's seat just behind it, while instead of the *front* elevating plane of the earlier models, the elevating surfaces were at the rear of the fixed tail plane. The Breguet was one of these progressive type biplanes of 1911. Constructed very largely of steel, it had a long, tapering body with its controlling planes—rudder and elevators—at the rear. Instead of a number of wooden supports between the planes the Breguet had exactly four reliable struts.

Henry Farman developed a military biplane in 1911 which had one particularly new feature. Instead of the upper main plane being placed exactly above the lower it had been moved slightly forward or "staggered"—giving it an overhang in front. The idea was that this gave a greater climbing power and was helpful in making descents, though the point has never been satisfactorily proved.

Until 1911 Germany had pinned her faith almost wholly to the Zeppelin as the unit for the aerial fleet which she had hoped to build up, and she had confidently expected it to prove its superiority to the heavier-than-air machine in the event of war. No funds had been spared to rush the work of designing and constructing these huge air monsters. Carefully and quietly the perfecting and standardizing of the Zeppelin under government supervision had moved forward, and German engineers had not been behindhand in designing engines particularly suitable to aircraft. While France was amusing herself with the clever little monoplanes and biplanes of the pioneer days— machines which could fly but a few yards at low altitude, Germany, possibly with the dream of world conquest tucked away in her mind, was sparing no expense to get ready her fleet of lighter-than-air craft. Imagine her chagrin when the feeble winged birds of 1908 and 1909 became the soaring eaglets of 1911, swiftly circling the sky, swooping, climbing and performing aerial tricks which made the larger and clumsier Zeppelin appear as agile as a waddling duck.

Whatever the feelings of the German military authorities were on the subject, they wasted no time in crying over spilt milk, but at once began a policy of construction by which they hoped soon to outstrip their brainier French neighbours. As in everything German, *method* was the characterizing feature of the airplane program they instituted. France had sought to encourage makers of all types of planes, and thus obtain a diversity of machines of wide capabilities. The plan did not appeal to Germany. From the very beginning she aimed at reducing everything to a fixed standard and then turning out airplanes in large numbers. When the war broke out it seemed for a time that she had been right, but it was not long before she looked with sorrow upon the sad lack of versatility of her fleet of standardized biplanes. They were hopelessly outdistanced and outmanoeuvred by the small, fast fighting machines of the French, while they were by no means so strong as the heavy service planes the French could put into the air.

Italy, Austria, Russia, America and Japan began also to make plans for the building of aerial fleets about 1911. The Italian Government relied at first on machines secured from France, or on those copied from French designs. Soon her own clever engineers began to be heard from and she was responsible for developing several of the powerful modern types. Russia would scarcely seem a country where aerial progress might be expected, yet she has given a good account of herself in aviation, and one of her machines, the giant *Sikorsky* did splendid work on the several fronts during the war.

I. I. Sikorsky, the inventor of the big Sikorsky machine was a little while ago merely a clever student at the Kieff Polytechnic. Like many other young men he dreamed of aerial conquest, but received little encouragement in carrying out his projects. At twenty-four, however, he became a student aviator, and almost immediately began work on original airplane designs. He succeeded in building a small monoplane which in some ways resembled the Blériot, except in its habits of flight. In these it was quite balky, refusing to fly except in short hops and jumps. Sikorsky's friends good-naturedly nicknamed it *The Hopper*.

But the young student was not one wit daunted. He plugged along steadily at new designs, and in the autumn of 1910 he actually took to the air in a tractor biplane of his own construction. Several other machines of somewhat the same type followed, and his efforts finally won the attention of the great Russo-Baltic Works. They offered him financial assistance to carry on his study of the airplane problem. With this backing Sikorsky moved forward to sure success. In the meantime he had secretly prepared plans for an enormous airplane which at first he dared not divulge for fear of ridicule and disappointment. Finally he took courage and laid them before his friends at the Russo-Baltic Works. Whatever they may have thought of his wild scheme of air supremacy they consented to give it a tryout, and in the Spring of 1913 the first of the giant "Sikorsky" machines stood awaiting a flight. It was viewed with grave misgivings by a number of experts, but to their frank surprise it took to the air with ease and flew well. The sight was a strangely impressive one. In wing span the big machine measured almost 92 feet, while the body or *fuselage* was over 62 feet long. The weight of the amazing monster flying machine was 4 tons. In the forward part of the fuselage cabins had been fitted, with a small deck on the bow. The fuselage construction was of wood, with a strong 8-wheeled landing chassis beneath it. Four 100 h. p. German "Argus" engines, driving four tractor propellers sent it racing triumphantly through the air. Its weight lifting ability was enormous, and it made a world record for flight.

Prodigious as this first great master of the air had seemed it was followed in 1913 by one still larger. The new machine was to the fullest extent an aerial wonder. Its enormous body consisted of a wooden framework covered with canvas, and in its interior a series of cabins were provided. There were three decks: the main one in the centre of the fuselage, designed to carry heavy armament of machine guns and a searchlight; a small deck at the stern; and one set in the undercarriage, where additional heavy armament could be placed. Only a few months before the storm of war broke over Europe this Air Leviathan was born, and at the time no one suspected it would so soon be called into active service. In the Spring of 1914 it made flight after flight, scoring a succession of triumphs by its record breaking performances, and winning for its designer a decoration from the emperor.

Sikorsky was a man of wealth but so recklessly did he lavish his personal funds on his airplane ventures that on many occasions he came very near to want as a result. It was no unusual thing to see him during those years of reckless experiment, braving the bitter winter weather of Russia in threadbare garments, shivering, but grimly and sternly determined. Then came the war, and at the first call his machines were ready to prove themselves in the battle against the Hun.

CHAPTER 4

The Airplane in the World War

Picture to yourself a scene outside one of the Allied hangars or airplane sheds, just back of the front lines, while the Great War is in progress. It is early morning, gray and chilly. Small fighting machines, which their trusty mechanics have carefully gone over for the tiniest flaw, now stand ready to take to the air. Pilots, wrapped in their heavy clothing—leather jacket, helmet and overcoat, gloves, goggles and muffler—prepare to face the frigid atmosphere above the clouds. The whirr of the motor, a short run over the ground, and up they go, one by one, until they become so many blackbirds, driving and looping and skimming through the sky.

Over in this corner is a large reconnaissance machine, with pilot and observer, waiting to ascend. It is one of a squadron that will fly over the German lines to take photographs of the enemy's positions. With its rapid-firing machine guns it is prepared to give battle to the swifter enemy craft that will flash out to challenge its onward flight. Its *rôle* is a difficult one. It cannot climb to safety as the fighting machine can do and then swoop down on its enemy from a favorable height. Its duty is to bring back accurate views of the territory on the other side of No Man's Land. No matter what the dangers, it must fly straight on, sticking close enough to earth to accommodate its camera's range, and deviating as little as possible from its course, though the enemy's speed scouts blacken the air with bullets and the anti-aircraft guns spit at it maliciously from below. All the machines in the squadron may not return, and there will be vacant chairs at the dinner table tonight when those pilots who have braved the stern hardships of the day relate their little experiences with the Hun. But those who do come back will bring information which will enable the Allied commanders to plan with intelligence the next move in the battle that is raging.

A tour of inspection would disclose still other machines, large and small, each designed and equipped for its special duties over the lines. There are heavy, slower-flying day "bombers," and—silent this morning but waiting patiently for the curtain of night to descend,—enormous night bombing machines, the fiercest and hugest of all the great birds of the flying force. Tonight, under cover of darkness these machines will speed upon their way, far over the enemy's lines. They carry fuel for a journey of many hours' duration, and heavy bombs which they will drop upon railway junctions, ammunition factories, staff headquarters and important positions deep in the territory of the Hun. Before they turn their noses homeward they will have crossed over the borders of Germany, and along their silent course fires will shoot up and enemy supplies and storehouses will be smouldering ruins when day breaks.

Unlike the night bombing machines of the Germans these great Allied aircraft will not drop their missiles upon open towns along the Rhine, nor will they leave behind them any toll of little children and civilians maimed and killed by their brutality. Their instructions are to bomb military objectives only, and when they have done that they will fly back silently through the night, passing over quiet villages and towns, where the sleeping inhabitants never will know that the great blackbirds have hovered so close to them.

When the war broke out airplanes were not planned so carefully nor equipped so fully for their special duties as they are today. Nobody foresaw exactly what those duties would be, and nobody once dreamed that the battalions of the air would play the tremendous *rôle* they have played in deciding the great struggle. Even Germany, who had been secretly planning and working and preparing for so long, had very little conception of the actual importance of her heavier-than-air machines. She neglected to use them entirely when she began her swift stride across Belgium. That piece of neglect lost her the prize, for the plucky Belgians, seizing the opportunity, marshalled their air forces, a small handful of airplanes, and used them to good advantage in discovering the intentions of the enemy. By means of her air force, Belgium was enabled to hold back for awhile the onrushing tide of the Hun armies, until France could bring her men into the field and the "contemptible little army" of Britain could be hurried across the Channel.

As the air forces were the deciding factor in that first great onslaught, so they have remained during the whole struggle. They began as mere scouting machines, but they have taken upon themselves more and more duties, until at the present time they are used for a multitude of purposes, and are fitted with the most perfect equipment to carry out their various ends.

Airplanes have often been called the "eyes of the army," but in war it is not sufficient to be able to *see* what the enemy is doing or is about to do. You must also be able to keep him from knowing what your plans are. So, there are the machines whose duty is to "see" and those whose duty is to "put out the eyes of the enemy." These latter must keep an eternal vigilance over the lines, on the lookout for enemy craft. When one is spotted they dash out after it, pursue it back to its lines and prevent it from performing its mission of reconnaissance. Nor are they satisfied merely to drive it off, they follow and give fight. Over there against the sky you see a little puff of smoke and flame that goes shooting down to the horizon. It is an enemy plane that will never again come spying upon Allied troops. Perhaps a group of fast German fighting machines dart out unexpectedly to avenge it, and then there is a terrible battle in the clouds, with every machine that is in the air hurrying to the skirmish. You try to follow their swift movements as they loop and dart and dive, but all you can see is a rapid confusion of wings, and now and then a machine that separates itself from the general *mêlée* and goes crashing to earth.

Not the least dangerous of the many services the airplane is performing is that of the artillery "spotter." It belongs to some particular battery whose guns are thundering away at the enemy. Hovering above No Man's Land, where its position is a trifle too exposed to be comfortable, it radiographs back to the gunners the exact locations of important objectives, then watches the firing and reports the results. Thanks to it the big guns do not speak in vain, and almost every shot is a direct "hit."

And then there are the dreadnaughts of the sky who actually take part in an attack, flying low over the lines and attacking the enemy infantry with guns and with death-dealing bombs. They must run the gauntlet of the enemy's fire, but on the other hand they spread terror and confusion in the ranks of the soldiers massed below, distracting their attention and leaving them open to the surprise of a sudden onslaught of Allied troops.

There are other machines which help in an attack by keeping the various parts of the long line in close communication with each other, so that all efforts are in unison. Their duties correspond in a way to those of the swift horseback rider we read of in the stories of old wars, who sped with news of great import from one commander to another. Only that the airplanes of today are so much more efficient than the gallant horseback rider of old, that although the line stretches across a nation, it can act as a man when the moment comes for a big "push."

Long before the war Germany had been busy turning out airplanes in large numbers in her factories, and in August, 1914, her air force was far superior in numbers to that of her great opponent France. She fondly imagined that she would be able with the greatest ease to put out her enemy's eyes, but in this she failed utterly. In spite of her military program of construction, according to which airplanes were turned out as if by clock-work, there was something wrong with her calculations. It is amusing to look back and see how German "method" had been carried to the absurd point of defeating itself. In manner truly characteristic, the Hun had standardized his airplane down to the last bolt.

Every machine turned out was of exactly the same pattern, and built up of exactly the same parts—parts which could be manufactured in large quantities and put together with unusual speed. It was certainly *system* raised to the *n*th degree. And the machines themselves were good enough—sturdy biplanes intended to be maids-of-all-work over the front lines. Yet in a little while after the fighting had begun, Germany withdrew them in more or less chagrin, and set herself to constructing others of varied patterns. They were well made and splendidly equipped, but they were not sufficiently *specialized* for the many different kinds of work they were called on to perform.

France had a motley array of airplanes of every size, shape and make when the war broke out. They had varying systems of control, so that a pilot who flew one with ease was nothing more than a novice when he stepped into another. He did not know how its new set of levers operated, nor how the plane would behave in the air. Moreover, the parts for these French airplanes and for their engines had been specially designed by each maker, and were quite unsuitable for any other type of machine. The result was that when a machine had to be repaired at the front, it was "laid up" for a long time, while the special part it required was being ordered and made for it. When finally it arrived, very often there had been some mistake, and so there was another long period of uselessness.

France had prided herself on her versatility in airship design. She now had cause to regret it as she viewed the almost helpless confusion it had caused in her air service. Her machines, moreover, were much inferior to the German in armament, speed and climbing gauges, cameras, and all the hundreds of accessories which gave the German machines their initial advantage. But experience is the best teacher, and no sooner had she seen wherein she fell short than dauntless France mustered all her resources to correcting past mistakes. Order was brought out of confusion, and it was only a very little while before the German war lords had need to look to their laurels, for the Frenchmen were far outstripping them in the air.

There was one "accessory" which the airplane of the Hun lacked, and which all his mechanical skill and ingenuity were not able to provide: *a pilot with the dash and daring of the French!* Even in those first dark days when the French planes were the equals of their adversaries neither in numbers nor in capabilities,—a continuous stream of gallant French pilots took to the air and proved that they could surprise and outmanoeuvre their slower-thinking opponents. While they held the line in their inferior craft, French manufacturers were rushing newer and better equipped machines to re-enforce them.

Great Britain was far behindhand in aircraft production when the trumpet of war sounded,—in fact, her air force was considered a negligible quantity by friend and foe alike. By dint of persevering search she managed to scrape up a small group of planes of many makes and for the most part antiquated. She sent them—along with her "contemptible little army"—to France, and there they succeeded in holding their own during the first great German push. When the stories of heroic fighting against hopeless odds, of British airmen flinging their lives in challenge against the foe in the great air struggle, began to reach home, the British lion repented his tardiness and a program of aircraft construction on a large scale was instituted without delay.

In carefully standardizing those first airplanes there was one point which the crafty Germans overlooked: which is, that you can't make a dray horse run fast, nor a race horse draw heavy burdens. The same thing holds good with the "steeds" of the air. A plane which is designed for great speed is never as good a burden bearer as one which is built to lift heavy weights at the expense of swiftness in flight. As soon as the duties of the airplane began to be specialized, the airplane itself began to appear in certain definite types.

Now of course the duties of the airplane in wartime are numberless, but out of the early confusion *three* types of machines were finally evolved, which, with the addition of equipment, such as a camera, machine guns, etc., are suitable for practically any sort of work over the land. They are:

1. The high speed fighting machines.

2. The reconnaissance machines.

3. The bombing machines (including the day and the night bombers).

Of all military airplanes there is none so fond of "aliases" as the high speed fighting machine. Possibly in order to baffle the uninitiated, or to surround itself with an atmosphere of uncertainty and romance, it goes by first one title and then another. Most often we hear it called a *speed scout*, perhaps for the reason that *it does no scouting!* At other times it masquerades proudly under the fine French titles of *"Avions de Chasse"* or *"Avions de Combat."* It is referred to as a "chaser," a "pursuit machine," a "battle plane" and a "combat machine"—but whatever it is *called*, in type it is the small, fast airplane, usually a single seater, quick in climbing, agile as an acrobat, able to "go" high and far,—for its duty is to run every enemy machine out of the sky and sweep the board clean before the heavier service machines begin their tasks of the day. It should be able to reach a height of from 18,000 to 23,000 feet, or in the language of the air, it must have a high "ceiling." From altitudes so tremendous that they awe the mere earthly pedestrian it swoops down upon its unsuspecting victim, opening upon him a stream of machine gun fire. For its pilot is also a skilled gunner and a crack shot. Upon his ability to manoeuvre his machine swiftly and cleverly and hit his target unerringly depends his own life and the life of a costly military airplane.

The reconnaissance machines and the bombing planes may do valuable service,—and indeed they invariably do—but it is the "speed scout" that covers itself with glory. The reason is that its career brings it nearer to the "personal combat" of the knights of old than anything in modern warfare. Driving his swift Nieuport scout as a knight would have ridden his charger, the beloved Guynemer[1] went forth to challenge the German fighters,—and other Frenchmen and Englishmen and Americans have followed him. It is a fact beyond all question that this branch of the service has produced some of the most truly

---

1. *Guynemer: Chevalier of the Air* by Henry Bordeaux & Mary R. Parkman *Georges Guynemer, Knight of the Air* by Henry Bordeaux and *The Chevalier of Flight: Captain Guynemer* by Mary R. Parkman also published by Leonaur.

unselfish and heroic figures of the whole war. The "speed scout" pilot did not need to be a man deeply versed in military affairs—as for instance the pilot and observer of the reconnaissance machine must be,—but he did need dauntless courage, unfailing nerves of steel, dash and daring and contempt for his own safety. So wherever the "speed scout" has blazed its trail of fire across the sky, there have sprung up the names of men whose heroic deeds have made them the idols of the whole world. Usually they have been very young men—young enough for their ideals to have kept fresh and untarnished from the sordid things of life, and thus they have written their names among the immortals.

Less appealing to the imagination, perhaps, but no less vital to the progress of modern warfare, is the slower flying reconnaissance craft. This machine is always a two-seater, and sometimes a three, for at the very minimum it must carry a pilot and an observer, while a gunner is a very convenient third party in case of an attack from enemy scouts. This type of machine is used for photographic work, for artillery "spotting," and for many general service duties over the lines. In the early days of the war it was customary for the photography airplane to be escorted on its mission by a group of fighting machines, who hovered about it and engaged in battle any airplanes of the enemy that might seek to interrupt its important work. But the last year or so have brought many improvements in airplane construction and it has been found possible to build a machine which can not only carry the heavy photographic apparatus and a couple of machine guns, but which can also travel at a good speed and climb fast enough to escape from the anti-aircraft guns. Instead of the rather helpless, clumsy, slow-flying reconnaissance machines of the early part of the war, we now have powerful "aerial dreadnaughts," which no longer need to run away, but can stay and fight it out when they are interrupted in the course of their air duties.

Military photography is one of the most fascinating of the side issues of the war. Before the day of the airplane it was the scout or spy who worked his way secretly into the enemy's lines and at great personal risk,—and often after many thrilling adventures, if the story books are to be believed—brought back to his commanding officer news of the disposition of troops, etc., in the opposing camp. Today the spy's job has been taken away from him. No longer is it necessary for him to creep under cover of night past the guard posts of the enemy. A big, comfortable and efficient airplane flies over the ground

A PHOTOGRAPH OF NORTHERN FRANCE TAKEN AT A HEIGHT OF THREE THOUSAND FEET.

AN AIRPLANE VIEW OF THE CITY OF RHEIMS, SHOWING THE CATHEDRAL

by broad daylight and collects the necessary information a great deal better than the spy ever could have secured it.

A reconnaissance camera has very little in common with a Kodak. The observer does not tilt it over the edge of the machine, focus it on some interesting object and "snap" his picture. As a matter of fact it works more after the manner of a gun. It is fixed in the bottom of the airplane, facing downward. The observer has been instructed before leaving the ground that a certain area or trench is to be photographed. Straight to the beginning of that trench line the pilot heads his machine. The observer compares the country over which he is flying with the chart or map which he carries. Just as a gunner sights a target, he locates the beginning of the trench line to be photographed through a bull's eye, and immediately pushes the button which sets the camera working. From that point the camera operates automatically, taking a series of overlapping pictures of the country it looks down upon.

With calm determination the pilot holds his machine to the course laid out, in spite of any opposition that may arise in his path, for the slightest deviation from that fixed line of flight will mean a gap in the reconnaissance report which the pictures represent. But once he has covered the required area, he turns and flees. In less time than it takes to tell that magazine of films is being developed in a dark room. From there the printed pictures are rushed to an expert interpreter who reads the secret meanings of the things he sees—this or that dark blotch or peculiar looking speck suggests to his trained mind a machine gun nest, a railroad centre, an observation post, a barbed wire entanglement, a camouflaged battery, an ammunition dump, or whatnot. Pasted together so that they give a continuous view of the foe's territory, the printed pictures are hurried to headquarters, where in a few brief moments their message has been turned into a command to the troops. By the word that those pictures bring the battle is directed, and the blow is aimed straight at the enemy's vital spots.

Occasionally instead of a series of photographs of a trench line or limited area, a continuous set of pictures of a broad space of country is desired. Then instead of a single machine as described above, a squadron of reconnaissance machines set forth, flying in V formation, with the leader of the squadron flying in front at the point of the V. The moment he reaches the area to be photographed, he notifies the machines behind him by firing a smoke rocket with a signal pistol. At that signal the V broadens instantly, so that it becomes almost a straight line, the commander keeping only slightly ahead so that he may lead the way.

On and on that broad V formation of airplanes sweeps, every camera registering, and all keeping close enough together to produce slightly overlapping photographs. Each machine will bring home a long line of pictures of the country over which it passed, and those lines, pieced together, will make a large military map of the entire region. That is if everything goes smoothly, which in war time it seldom does. More likely that plucky V will be pounced upon by a herd of fast fighting machines whose duty it is to see that none ever return with their information to headquarters. There will follow a terrific contest; the observer in the reconnaissance machine becomes a gunner, and fires away at his pursuers, while the never-failing camera keeps steadily on with its job of recording.

As nearly as possible the V formation is held, for much depends upon it, but suddenly a great gap appears in the line. "Done for" with a direct hit, one brave machine goes crashing earthward. That will mean a gap in the "map" that is in the making. Still the V presses on relentlessly. One of the planes begins to lag behind. There is something wrong with its engine. It does its best to keep up with its fellows, but soon it is left behind, and the enemy craft dive after it. Battered and torn, its numbers depleted sadly, several of its crew wounded, its wings perhaps riddled with bullets, the photographing squadron turns its face toward home, and, flying now as high as possible to keep out of sight, puts on all speed for the safe side of No Man's Land. Military photography *sounds* easy and comfortable. It *demands* the type of courage which can make a man stick to a given line of flight, even when certain death lies straight ahead.

Sometimes a machine carries both bombs and a camera, and, as it drops its missiles, keeps a continuous record of its "hits" to carry home. And that brings us to the bombing machine, last but not least of the trio of military airplanes.

The bomber that works by day and near to its own lines, is similar to the reconnaissance machine, except that it does not usually carry a radio apparatus or a camera. Instead, the greater part of its cargo consists of bombs, dread instruments of destruction which will fall on the railroad junctions, troop trains, staff headquarters or ammunition dumps of the enemy. The day bomber is never used for long distance work, and so it does not need to be of tremendous size, as the machine which must carry fuel for an all night run as well as a large quantity of bombs to drop on a far away important objective.

The night bomber is the giant of the sky. The greatest genius of the cleverest designers has been expended upon its construction. More and more its tremendous importance is being recognized. Its activities precede every great offensive movement, for it flies over the enemy's country, leaving a trail of terrible destruction in its wake, and "preparing the soil" for the infantry advance. Deep in the territory of the foe it searches out the great supply centres and railway terminals and there it unloads its cargo of bombs.

If the Allies had possessed a sufficient number of these huge bombing planes they could have carried on an aerial warfare against Germany which would have defeated her without nearly so great a sacrifice of the lives of the infantry. The work is dangerous, but a single bombing plane could have wreaked more vengeance upon the Hun than perhaps a whole regiment of the bravest fighters. Consequently its use would have meant economy of human lives.

These fearful shadows that walk by night require pilots of the utmost skill to navigate the sea of darkness, as well as bomb droppers and gunners whose training has been perfect. The largest of them are equipped with either two or three powerful engines, each working a separate propeller. Such a machine can carry as much as five tons of explosives, with fuel for a twelve hours' flight.

The night bomber is very often a huge triplane, for the extra wing surface gives greater lifting power. At the same time the triplane has greater stability and has a fair chance of reaching home even when one of its planes has been badly damaged. It is the same with a machine which has two or more engines: even when one of these has been put out of order by the shots of the enemy the airplane can still reach home. The night bombers must travel long distances, carry great cargos, bomb their objectives and make their escape, and so in the construction of their machines as much stability, lifting power and speed as possible has been the aim.

Usually it is some important munition base or factory centre that is supplying the German troops, which the airmen set out to bomb. They travel in squadrons not only for safety, but because in this way an almost unlimited number of bombs can be carried and dropped simultaneously. Often a second squadron follows the first at a short distance. By the light of the terrible fires that the first set of explosives dropped are bound to start, this second squadron can drop its bombs with greater precision directly on important buildings that must be destroyed.

Moving slowly under their great load of explosives, and flying low, these two squadrons of destroyers start for some point in the heart of the German Empire. Like ghosts they "feel their way" mile after mile. They are not anxious to invite detection, for under the great weight of their "messages" to Germany, they would not be able to manoeuvre quickly or to climb to safety.

Once those tons of explosives have been released and the noise of their dreadful havoc has aroused the anti-aircraft gunners of the enemy, those bombing planes will find the earth an uninviting sort of region and they will be glad to spring into the protecting silence and darkness of the upper air. And this they can do easily, for, rid of their load, they possess unusual climbing powers. The second squadron of bombers, flying over the same territory may expect a warm reception, and they will need to do their work quickly and beat a hasty retreat.

Such are the mysterious doings of the night. When the early dawn appears, gray and heavy eyed, it will find the bombing planes tucked away drowsily in their hangars, scarcely knowing themselves whether the journey up the Rhine was a reality or merely a terrifying dream.

And with the dawn their daylight sisters will take up the work near home. Word has just come that enemy reinforcements are moving up to the front along certain roads. "Fine," sings out a young lieutenant, appearing unexpectedly on the field from a small, carefully camou-flaged office. "We will make them dance for us this morning!" He talks quickly and determinedly with a group of pilots, giving instructions, charging all to keep the formation. Machines are gone over to make sure that everything is in perfect condition. Then the first bombing plane, bearing the flight leader, "taxis" across the field, appearing to stagger under its great burden. Suddenly it takes to the air, and like a large graceful bird, its clumsiness all gone, it soars up into the blue. Rapidly the other big birds follow suit, and at a signal they are off, the flight commander heading the group, and the others following in close formation, like a huge flock of wild geese.

On and on they fly, until beneath them appears the winding rib-bon of road that is their objective. It is crowded with marching troops, gun wagons, supplies. As they swoop close to the earth they catch a swift glimpse of white faces turned up at them with terror. Then panic falls upon the marching column and, helter-skelter, every man tries to break away to a point of safety. In another moment guns are turned upon the bombers, but they dodge the flying shells and let go their heavy explosives, which crash to earth with dreadful uproar. Where a few moments before the Huns were following their way undisturbed there is now a road in which great furrows are ploughed; huge holes gape open and a hopeless mass of *débris* covers the earth. The columns of the enemy will be blocked for many hours while the mass is being cleared away. Satisfied with the results of their exploit the bombing squadron turns swiftly toward home.

How simple a matter it seems at first glance to release a bomb and hit a given point below. Actually it requires the very highest skill. To begin with, the airplane is moving at tremendous speed, and the bombardier (as the man who drops the bombs is called) has to know exactly how the forward motion of the airplane will affect the direction that the bomb takes on its course toward the earth. Moreover the bomb has a speed at starting equal to the speed of the airplane, and this beginning speed is increased by the action of gravity drawing it down. It may be aided in its journey by the wind or retarded, according to the wind's direction, and this too must be taken into account, if the target is to be hit. Bomb dropping can only be carried out successfully with the aid of the most delicate and complicated range-finding mechanism, with which every bombing plane is equipped. The Germans have led the way in inventions for this purpose, and their Goertz range finder is perhaps the best in the world. (As at time of first publication).

The bombs themselves are generally carried in vertical position, one-above another, in the body of the airplane, and by an automatic arrangement, as one is released, another slips into place, ready to be dropped.

Now that we have made the acquaintance of the three types of machines that are used over the trenches—the "speed scout" or small fighting machine; the larger armed reconnaissance plane; and largest of all, the bomber—let us go back and give just a hasty glance at the main points of their construction.

First we must recall the "A B C facts" we learned about wing construction. A wing gains lifting power from two sources: the upward pressure of the air current underneath it, and the force of the vacuum above it which is created by the arch of the wing. If a wing is only slightly arched it can move *forward* through the air more swiftly, but it will not have the *lifting power* of the high arched wing. This is the reason that an airplane which must be a weight carrier cannot be as fast in flight as the "speed scout," which has only its pilot and a machine gun to carry.

The "speed scout" is always a small machine, usually a single-seater, with a gun in front that fires over or through the propeller. In the early part of the Great War it was most often a monoplane, but the smaller biplane took its place, because, with practically the same speed, it combines greater stability.

The planes of the speed scout are very flat as compared with those of the reconnaissance craft. This airplane must carry machine guns, photography apparatus, radio, and a pilot, an observer, and often a gunner. Its wings must therefore be arched to give it lifting power, but at the same time it becomes a much slower flying machine than its smaller sister.

Lifting power of a wing can of course be increased *up to a certain point* by increasing the wing area, so that a greater air pressure is created below. Beyond that certain point the machine would become unwieldy and would lose its balancing properties. Yet this idea has been put into practice in building the latest types of aerial dreadnaughts used for reconnaissance. These airplanes have gained their lifting power partly by increasing the wing spread and partly by arching the wing. Thus a wing has been secured which offers the minimum resistance to forward motion through the air, together with the maximum weight carrying ability. Biplanes of this type are by far the most popular of those designed for general service, for they combine speed, climbing ability, and lifting power,—thanks to their strong armament they can defend themselves or run away quickly as the situation demands.

But there is one other method which has not yet been mentioned of increasing the lifting power of an airplane. It is simply to add a third wing. When we have made the wings of the biplane as large as we dare, and have curved them to make them weight-bearers, if the resulting machine is still not strong enough to carry as many tons of explosives as we desire, there is only one thing left to do and that is to add a third wing. Thus the triplane made its appearance in answer to the call for planes which could carry vast cargoes of explosives and fuel for journeys of many hours over the enemy's country. The huge night bombing machine of the present time is almost always of the triplane type.

CHAPTER 5

Some of the Problems the Inventors Had to Solve

Every American must feel a glow of pride when he stops to think that it was two of his fellow-countrymen, Wilbur and Orville Wright, who invented the airplane. But it is largely to France, our great ally and friend, that the credit must go for improving upon the invention of the Wrights, and making possible the wonderful aerial feats, the marvellous flights and accomplishments of the airplane of today. From the first day they saw an airplane flown, the French were wildly enthusiastic. They gave freely of their money and their encouragement to help the good cause along. French inventors attacked the problems of the heavier-than-air machine with a will, and their unfailing determination and refusal to accept defeat or failure made final victory inevitable.

But before we could have the powerful fighting machines, the big cross country fliers and the seaplanes of today, there were many difficulties of construction which had to be met and solved.

First of all the pioneer designer had to choose between the monoplane, the biplane and the triplane. The monoplane was light in weight and could fly faster with the same powered engine than the biplane. But it was difficult to know just how to brace and strengthen the single pair of wings. In the biplane the struts between the wings gave strength and firmness. The wings of the monoplane were braced by wires to the body, but often they did not prove strong enough and the airplane collapsed in mid-air. In spite of this danger the monoplane was much in favour because of its speed.

Slower in speed, but stronger and a better weight lifter was the biplane. And in addition to strength it possessed more natural stability, a much sought after quality in the pioneer days.

Even more stable and with greater lifting powers than the biplane was the triplane, but the difficulty here was the lack of an airplane motor of sufficient strength to drive it. Until clever engineers came to the rescue with an improved aircraft motor, the triplane was very much in disfavour.

The monoplane, indeed, captured most of the early records for speed and it was this type of machine that was generally built by the sportsman type of airman, while men like the Wright brothers and others whose aim was to develop an airplane of unusual reliability and suited to many purposes, turned to the biplane and gave many hours and months and years of their time to its improvement.

Once the choice of a *type* had been made, there were countless other problems. *Stability* was of prime importance and the airmen of a few years ago laboured desperately to attain it. They knew all too little about the airplane from a scientific angle. We have seen in our brief study that the method of obtaining balance in a glider or an airplane is to see that its *centre of weight* coincides with the centre of the *upward pressure* of air. How to bring this happy state of things about was a source of much debate. Some suggested that instead of a tail at the stern a tail in front of the main planes of the machine would help to balance it in flight. Some placed the pilot's seat above both planes of the biplane, while others thought he should sit below. Many of these queer ideas were tried out and by dint of hard practise and many failures certain simple elementary facts were finally weeded out and set down.

Probably the addition of a "fuselage" or body to the modern airplane has had something to do with helping in the proper distribution of its weight and increasing its stability. Larger at the bow and tapering toward the stern where a fixed tail piece or horizontal stabilizing plane is attached, it resembled more or less closely the general outlines of a fish or bird. And this "streamline form" greatly reduces the *head resistance*, another important subject on which there was very little known when the first of the airplanes was built. In addition to having only a very slow and inefficient engine the early machine suffered from the head resistance it created as it pushed forward through the air, and this check to its progress ate up the little speed its motor could develop. For if the airman of 1908 or 1909 was made miserable by his fear of winds, gusts and aerial whirlpools which might upset him in mid-air, his fears in this direction were completely overshadowed by his worries about a suitable motor. If the design of his craft was faulty and it proved "balky" when he attempted flight, he had only himself to blame. But for an engine he had to rely entirely upon someone else. The airplane could be a "home-made" article, but the engine had to be chosen from such as were on the market.

The Wright brothers in their first flying machine used a made-over automobile engine of 12 horsepower. It was not long before this was improved upon, and later Wright machines had a four-cylinder, water-cooled engine developing 35 horsepower. Its weight had been reduced as far as possible and its simplicity of design was its greatest recommendation.

Undoubtedly the engine problem has been the big one in the history of aviation. The coming of the internal combustion engine might be said to have placed practical aviation within the range of possibility, but at that it took a long time to evolve a motor especially suited to the needs of aircraft. There were three things needed in an airplane motor: *Light weight, high power,* and *absolute reliability.* How important the third factor is we can imagine if we stop to think that nothing keeps the heavier-than-air machine afloat but its own speed, creating an air pressure beneath its wings. Like the boy who runs with his kite in order to make it go up, the airplane must "go" if it would rise, and the moment its engine fails there is nothing to prevent it from falling to the earth.

The driver of a motor car, can, if his engine goes wrong, get out and go over it carefully until he finds what the difficulty is. The pilot of an airplane, soaring thousands of feet above the earth, is at the mercy of his motor's reliability or lack of it. Engine failure was, and still is, one of the greatest dangers the airman has to fear. Another chief cause of trouble in early airplane motors was overheating. Before actual airplane engines had been designed there was nothing to do but to use the type of engine which had been designed for the automobile, with as much reduction in weight as could be secured. But the automobile engine was never intended to run at top speed continuously and for long periods, as the airplane engine necessarily must do. In a car the motor has little stops and rests, as it is throttled down for a moment or changes in speed are made, and these breathing spells help it very much indeed in the "cooling off" process. The airplane engine does not have these little between-time naps. The result was that the automobile engine installed in the early airplane invariably overheated and caused serious trouble. Under these conditions no flights of any distance could possibly be attempted.

Yet at the Rheims Meeting of 1909 Henry Farman surprised the world by remaining in the air two hours in a continuous flight. Up to that time the feat had never been equalled or approached. Aviators were amazed and sought an explanation. The answer was: the Gnome motor.

Anxious to help the airplane in its forward march, French engineers had good naturedly set to work and the Gnome motor was their first answer to the anxious question of "What engine?" It involved a new and ingenious system of cooling which made it possible for Farman to drive his big machine round and round the Rheims course until stopped by darkness, but without ever experiencing the slightest difficulty with his motor.

Before attempting to understand the secret of superiority of this first real airplane motor over others of its day, we must know a little more about the elementary principles of any internal combustion engine. The diagram shows *one cylinder* of such an engine in action.

A mixture of gasoline and air—called "carburetted air"—is introduced through a valve opening into a chamber or cylinder, as shown in figure A of the diagram. The valve opening then closes, and the piston moves forward compressing the gases enclosed in the cylinder, as shown in figure B. An electric spark suddenly explodes these compressed gases, causing them to expand with the greatest violence and drive the piston back. This action, which is shown in figure C, is called the "power stroke," for, transmitted by the piston rod to the crankshaft it furnishes the power which turns the propeller and sends the airplane forward through the air. Just before the piston reaches the end of the power stroke the exhaust valve opens, and the exploded gases are forced out of the chamber, partly by the force of their own tension and partly by the upward stroke of the piston, as shown in figure D.

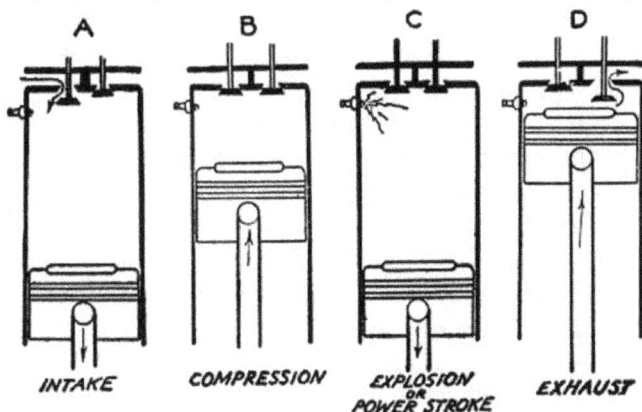

DIAGRAM OF AN INTERNAL COMBUSTION ENGINE CYLINDER, SHOWING PRINCIPLE ON WHICH IT WORKS

The carburetted air is supplied to the cylinder from a chamber called the "carburettor." Here it is produced by the mixture of a gasoline spray—similar to the fine spray of an atomizer—with the air.

A spark plug is fitted to the cylinder, and a break current from an electric magneto causes the spark which at the proper instant explodes the compressed gases.

Since by means of the explosion of the gases the force is produced which drives the airplane propeller, the violence and frequency of these explosions determine the power of the engine. Greater power can be obtained either by increasing the size of the cylinder so that it can hold more of the carburetted air, making a greater explosion possible; or else by causing more frequent explosions. The latter is the better method in an airplane engine, as larger cylinders mean more weight to be carried. In the average airplane engine from 1500 to 2000 explosions or revolutions occur per minute.

The combustion cylinder of an aircraft engine is usually built of steel, and the piston of cast iron or aluminum, which furnishes a very smooth gliding surface. The piston rod transmits the power to the crankshaft, a long rotating piece of steel. Every time the piston rod is thrust down by the explosion in the cylinder, its motion serves to turn the crankshaft and thus the vertical motion of the piston is transformed into the rotary motion which sends the propeller whirling through the air.

Wherever two surfaces of metal must rub against each other, as in the case of the piston and the cylinder, there is bound to be a great amount of friction. This friction causes the parts to heat and in time it wears away the surfaces and destroys the efficiency of the engine. In order to avoid this, the surfaces must be kept constantly well oiled or "lubricated." In some engines all the parts are enclosed in one large box or "crank case" which is filled with oil. Small holes are bored through to the surfaces to be lubricated, and the oil is splashed upon them by the motions of the piston rod, the crankshaft, etc., as they plunge through the oil bath.

But overheating of the cylinder may cause this oil to decompose and in order to prevent this a "cooling system" is necessary. For only when the engine is kept cool and properly oiled can it be expected to run smoothly or give satisfactory service.

So now we come back to the problem of cooling, which caused so much anxiety and trouble to the early aviators. With their engines running at the great speed which was necessary to keep the airplane in the air, overheating and engine difficulties were sure to arise. Cooling of the cylinder is accomplished in one of two ways: either by water or by air. If water is used, a "jacket" in which the water circulates is placed around the cylinder,—the water as it becomes heated passing out of the jacket to the radiator, where it is cooled before it returns. The radiator, at the very front of the airplane body, is exposed to the swift current of the air as the machine drives forward, and this air current serves to reduce the temperature of the water.

This method was the one originally employed with the automobile engine, but in the early models the cooling system, though adequate for the motor car, was hopelessly insufficient when the same engine was installed in an airplane.

It was the Gnome manufacturers who first thought of a most ingenious scheme for cooling the cylinders of the internal combustion engine. Instead of having the piston and the crankshaft move, it was the cylinder itself which moved in the Gnome motor, while the crankshaft and piston were stationary. Thus cooling was very easily accomplished, for the cylinders, flying through the air, making as many as 1500 revolutions per minute, cooled themselves.

The crankshaft in the Gnome motor had been hollowed out to form a tube or pipe, through which the fuel or carburetted air passed to the cylinder by means of a valve in the head of the piston which worked automatically. The Gnome could be built up of any number of cylinders, according to the power required. Its cylinders were set in a circle about the crankshaft, so that the entire engine occupied a minimum of space in the airplane body. Scouted at first as a freak engine, it soon proved its superiority over all those in use and was rapidly adopted by builders of all types of airplanes.

Today the stationary engine has been greatly improved, its provisions for cooling have been increased and it is once more looked on with favour by many manufacturers of aircraft.

The cylinders of an internal combustion engine can be grouped in one of three ways, and thus there are three main types of airplane engines we should be able to recognize. They are the *straight-line* engine, the *V-type*, and the *radial*. In the straight-line model four, six, or even a larger number of cylinders are placed in a row in one crank case. In the V-type of motor they are set instead in two lines, like a letter V; while in the radial type the cylinders form a circle around the central crankshaft. The radial motor may be stationary or its cylinders may revolve, in which case it becomes a rotary engine, as for instance, the Gnome.

Each of these types of motors has its peculiar advantages. The least "head resistance" is caused by a straight line engine, and this type also uses less fuel and oil. But it is usually heavier in weight, owing to the larger cooling system necessary and the longer crankshaft, and it takes up more room in the airplane fuselage than a motor of the compact radial type. The radial engine is very light in weight,—a big item in the airplane—but it consumes a large quantity of fuel and oil and besides produces a maximum "head resistance." The V-type motor is a compromise between the two,—lighter in weight than the straight-line, less wasteful of fuel and causing less "head resistance" than the radial.

The rotary engine, because of its appetite for fuel and oil is no longer used in airplanes which are intended for long distance flights, because here the weight of the extra fuel carried has to be considered. In short distance, high-speed machines it works well, but in the larger planes the vertical or V-type motor has been found to give greater satisfaction.

When we read of the enormous trouble the pioneers of aviation went to, in order to find an engine suitable to drive the propeller of the airplane, we cannot help wondering just how the revolving of the propeller sends the whole machine flying forward through the air. The matter is very simply explained. The propeller of a ship is often referred to as the ship's "screw," and though few people have ever compared it with the small screws they use about the house, or with the screw and screw driver in the tool chest, there is in fact very little difference in principle.

Take a screw and place it against a block of wood, and then commence to turn it with a screw driver. Straight into the wood its curved edges will cut their way, dragging the round steel rod of the screw behind them. With every turn they will cut in deeper and carry the screw forward through the wood. That is what the propeller of a ship or an airplane does: it screws its way through the water or the air. But of course there is this difference, that the wood offers great resistance to the forward motion of the screw, while the water offers much less resistance to the ship's propeller, and the air less still to the propeller of the airplane. If, as in the case of the screw-driver, the airplane propeller is in front of the airplane and drags its load along behind it, it is called a "tractor" propeller; but if instead it is placed at the stern of the airplane, and as it screws through the air it pushes the airplane along ahead of it, then it is known as a "pusher" propeller.

The little cutting edge that winds round and round an ordinary screw is referred to as its *thread*, and the distance between two of these edges or threads is known as the *pitch*. In some screws the threads are very close or, to put it another way, the pitch is small, while in others it is much greater. Each blade of a propeller is really a portion of a screw. To go back to the example of the screw-driver and the block of wood, every time the screw is turned once around it will advance into the wood a distance equal to its pitch. The same thing is theoretically true of the propeller of an airplane; at each revolution it might be expected to advance through the air a distance equal to the pitch that has been given to its blades.

But the air may allow the propeller to slip back and so lose some of its speed, a thing which was not possible with the screw-driver. This tendency to slip varies with the pitch of the propeller and the speed of its revolutions. A propeller which works splendidly when turning at a given rate, may prove worse than useless when the engine is slowed down and it is only making half the number of revolutions per minute. And so we begin to see another of the big problems of the pioneer airmen: to determine the right pitch for the propeller in relation to the speed which had been determined upon for the airplane. It is a problem that has not been wholly solved today, because of the fact that an airplane cannot always be driven at "top speed." If the maximum speed of the machine is 150 miles per hour, and the propeller has been designed to deal with the air efficiently at this speed, it is apt to slip and slide and waste away the power of the engine when for any reason it is necessary to slow down to 100 miles per hour. The only answer to the difficulty is a "variable pitch propeller" which may be altered to conform with alterations in speed, but up to the present time nothing really satisfactory along this line has been devised.

Another question in connection with the propeller has been of what material to make it. Wood is most generally used today, for although steel and aluminum have been tried, they have not been found to stand the strain so well. Imagine for one moment the stress upon an airplane propeller beating through the air at the rate of 1500 revolutions per minute. The greatest strength has been secured by building it up of several pieces of wood which are fastened strongly together and varnished.

Materials have always presented a source of endless experiment and differences of opinion in the construction of the airplane. The problem has come up in connection with the fuselage, the wings and wing coverings, the landing chassis—in fact, each and every part of the heavier-than-air machine has raised the old query: "What shall we make it of?"

In the earlier machines wood was almost entirely used in airplane construction. For one thing it was cheaper, and for another it was easier to get wood working machinery, than the complicated and expensive machinery necessary to construct airplanes out of metal. Metals are stronger but they cost more and they make the problem of repairs more difficult.

The wings of the airplane are usually built up on a wooden framework which gives them their shape and curve. Many have been the disputes over the matter of wing coverings. In the pioneer machines they were covered with cotton material which had not been treated to make it water-proof or air-proof. It gave the poorest kind of service, and an effort was made to improve it by rubberizing it, but this process did not produce a wing of lasting durability. Many other treatments were experimented with, but with little success until the substance known as "dope" made its appearance.

"Dope" is largely composed of acetyl cellulose. It makes the wing covering proof against rain, wind, and the oil thrown off from the airplane engine, and gives it a fine, smooth finish and excellent durability. Two or three coats of it are usually applied, with a final coat of varnish on top, to produce a wing that is sure to prove strong and trustworthy.

The problems of starting and landing the airplane have been many. The early Wright machine had to run on a little trolley down a track in order to gain sufficient momentum to take to the air. Later machines showed an improvement on this. Henry Farman attached wooden skids to the bottom of his airplane and fastened wheels to them by means of heavy rubber bands. Thus he could start his motor and run over the ground until his speed permitted him to rise, while in making a descent the wheels flew back on their flexible bands and the stout skids absorbed the shock of the fall. Most of the modern machines have a wheeled framework below the fuselage, which permits them to run over the ground in starting and also in making a descent. The danger of engine failure becomes very important when near to the ground, as the pilot has no time to get his machine into a gradual glide and avoid a bad accident.

This danger is sometimes averted by installing two engines, so that if one stops the other will carry the airplane on up into the air and prevent a smash-up. But the thing which has greatest effect on the ability of the airplane to land easily is its own design and speed. The wings of the airplane, its propeller and its whole construction have been planned so that it can support itself best in the air when flying at a certain fixed speed. Suppose this speed for a certain type of airplane to be 150 miles per hour. The airplane cannot land while travelling at that rate, yet its speed while still in the air can only be diminished to a certain point with safety, and below that point it may not be able to sustain itself in flight. The pilot must be able to land his machine without accident and without throttling his engine below this danger line; while the designer of airplanes must struggle to produce a machine which, while flying best at its maximum speed, will *fly* at a much lower rate of motion, when necessary to effect a landing.

The supporting power of the wings depends partly on their size and partly on their rate of motion. Small wings moving at high speed produce the same supporting pressure of air beneath them as large wings flying at slow speed. The problem of a safe landing could best be solved by building wings whose area could be altered in mid-air. When travelling under full power the pilot would reduce the wing spread, as the smaller wings would then be sufficient to support the weight of the machine and would create less air resistance. When about to land, he would increase the spread of the wings, so that at the slower rate of motion through the air he might take advantage of a larger supporting surface. Nothing of this sort has yet been worked out on a practical scale, but many have been the suggestions for "telescoping wings."

The reduction of "head resistance" and the "streamlining" of the airplane have received their goodly share of attention and experiment. Today the airplane fuselage is carefully streamlined, but the landing chassis beneath it creates a good deal of resistance to motion. Probably this problem will be solved by devising a landing chassis which, after the machine has arisen from the ground, can be drawn up inside the body, and let down again to make a landing, but this is another important question which is not yet worked out in the airplanes of the present time.

The coming of the war caused all nations to stop and take strict account of what had been accomplished in solving the many problems of aviation, for the war machine had to be as nearly as possible the sum total of all the best that had been worked out up to that time in the difficult matter. In aircraft design and in types of engines France undoubtedly stood foremost, although the knowledge she possessed had not been sorted, pigeonholed and accurately standardized as was the case in Germany.

Germany had some excellent aircraft motors of the water-cooled type, which were light in weight, very reliable and high-powered. The German government had spent large sums of money for the purpose of encouraging airplane construction and the improvement of designs and engines.

Yet no country at war found her military airplanes all she had expected them to be. It was not until actual war service brought definite demands from the pilots and definite criticisms of the bad features of the airplanes in use, that the designers were able to turn out machines of the highest efficiency.

There were many things which the pilots asked for. Speed and climbing power were among them, greater ease of operation, more protection in the way of guns and armament, the pilot's seat so located that his vision was not obstructed above or below, and a uniform system of controls. Gradually all these requirements have been met by the airplane makers. By 1917 they had turned out machines which could fly as fast as 150 miles per hour and climb to 22,000 feet, while since then even this record has been greatly improved upon.

In the field of aviation America can claim one big accomplishment since her entrance into the World War. That is the Liberty motor, probably the most successful motor that has ever yet been devised for an airplane.

When it was decided that we should begin work building American airplanes, there was one important problem: the engine. Foreign types of engines could not very well be built in this country, as they required workmen of many years' training in a highly specialized field. It was agreed that we must have a motor of our own, which could be manufactured rapidly under the conditions of our present industrial system.

Two of the most capable engineers in the country were summoned to Washington, and in order to assist them in their work motor manufacturers from all over the United States sent draftsmen and consulting engineers. For five days these two men did not leave the rooms that had been engaged for them at the capital.

Sacrifice was necessary if victory was to be won. Engineering companies and companies making motors for automobiles, etc., gave up their most carefully-guarded secrets in order to make the Liberty motor a success. The result was that an engine was produced so much better than anything on the market that our allies ordered it in large quantities for their own airplanes. Twenty-eight days after the drawings were started, the first motor was set up. It was ready on Independence Day, and was demonstrated in Washington. The parts had been manufactured in many factories, yet they were assembled without the slightest difficulty. The completed engine was sent to Washington by special train from the West. Thirty days later it had passed all tests and was officially the Liberty motor.

One of the most remarkable things about the Liberty motor is the way in which all of its parts have been carefully standardized so that they can be manufactured according to instructions by factories in all parts of the United States. The parts can then be rapidly assembled at a central point. The cylinders are exactly the same in every case, although the Liberty motor is made in four models, ranging from 4 to 12 cylinders. They can be interchanged and the parts of a wrecked engine can be used to repair another engine.

Thus American wit, patriotism and energy were able at a most critical time to answer the threat of German supremacy in the air. Our aircraft production has gone forward with speed which almost baffles understanding, while the airplane motors we shipped abroad in such overwhelming numbers to be installed in foreign machines gave good service to the cause for which the Liberty motor was named.

CHAPTER 6

Famous Allied Airplanes

Airplanes, like men, are not all alike, even when they are in the same line of work and performing the self-same duties. In war time, every gunner has his own little peculiarities, every sharpshooter has his personal ideas about catching the enemy napping, and every infantryman who goes over the top, in spite of his rigorous training in the art of war, meets and downs his opponent in a manner all his own. So it is with the machines that in the last few years have won fame for their valiant service over the dread region of battle. Roughly they can be lined up as fighting machines, reconnaissance airplanes and bombers. Yet if we look a little closer, individual types of planes will stand out of the general group, and it becomes fascinating to study them in their design, their achievements and their particular capabilities.

As it would be impossible to mention in one short chapter all the brave pilots who distinguished themselves for their heroism in the war in the air, so it would be a hopeless task to attempt to do justice to all the airplanes which rendered good service over the front lines. The best we can hope to do is to make the acquaintance of the most famous of them all.

There is one little machine, which, when the final retreat was sounded and accomplishments were reckoned, had covered itself with glory. Like the many famous pilots who have driven it, it has learned much by experience, and it has changed considerably in outward appearance since the summer of 1914. Wherever the achievements of the "speed scout" are mentioned the *Nieuport* is bound to come in for its share of the praise. This little fighting machine was greatly relied on by the French, who used it in large numbers over the front lines. Although lately another swift scout plane has come into the field to eclipse its reputation, it probably took part in more aerial battles than

THIS PHOTOGRAPH SHOWS THE RELATIVE SIZE OF THE GIANT CAPRONI BOMBING PLANE AND THE FRENCH BABY NIEUPORT, USED AS A SPEED SCOUT

any other airplane of the Great War.

It was the *Nieuport* monoplane whose speed and agility at manoeuvres made it a favourite in the early days of the hostilities. For a while it was a match for the German scout machines, but the rapid strides which aviation took under the pressure of war necessity left it behind, and the more rapid and efficient *Nieuport Biplane Scout* made its appearance. In several important features it was entirely different from any of the biplanes. It could not quite forget its monoplane construction, and it had made a compromise with the biplane by adding a very narrow lower wing. It was humorously nicknamed the "one and one-half plane," but it proved itself just the thing the fighting airmen were looking for. Its narrow lower plane, while giving more stability and a "girder formation" to its wing bracing, did not interfere with the pilot's range of vision, a highly important consideration. In order to allow as full a view as possible in all directions, it had only two V-shaped struts between the planes, while the upper wing, just above the pilot's seat, had been cut away in a semi-circle at the rear so that he might be able to see above. The lower wing was in two sections, one at each side of the fuselage.

This little biplane had a top wing span of only 23 feet, 6 inches, while the distance across the lower plane from tip to tip was a trifle shorter, measuring just 23 feet. The upper plane measured from the front to the rear edge a trifle less than 4 feet,—or to use technical language, it had a "chord" of 3 feet, 11 inches; while the chord of the lower wing was only a little over 2 feet. The entire length of the biplane from the tip of its nose to its tail was 18 feet, 6 inches. The fuselage was built with sides and bottom flat but the top rounded off. There was plenty of room for the pilot to move freely in his seat. Armed with a machine gun which fired over the propeller, he was well able to defend himself when enemy craft appeared.

The *Nieuport* biplane wrote its achievements in large letters during the Great War. It was the machine which Guynemer and his famous band of "Storks" flew in their daring battles against the German fast scout, the *Fokker*. It carried many an American chap to fame in the Lafayette Escadrille. England, Italy and America all used it over the lines, and its high speed and quickness at manoeuvre made it a general favourite. Today it is made in either the single-seater scout type, or in a larger, two-seated model. The gunner's seat in the latter is in front of the pilot, and a circular opening has been cut in the upper plane above him, so that in an aerial battle he may stand up, his head and shoulders

THE SPAD, THE PRIDE OF THE FRENCH AIR FLEET

above the top wing, and operate the machine gun, which fires across the propeller.

In spite of all its excellent qualities and its record of victories won, the *Nieuport* has lost its championship among the "Speed scouts." Another tiny biplane of still greater speed, has wrested the honours from it. The first place among fighters is now perhaps held by the *Spad*. Carrying one or two passengers and equipped with an engine of 150 to 250 horsepower, with its Lewis and Vickers machine guns spitting away at the enemy, it is a formidable object in the arena of war.

Not to be left behind, America has developed a small, fast fighting machine which bids fair to make the other two look to their laurels. It is a tiny *Curtiss triplane*, the span of whose wings is only 25 feet. Its extra lifting surface gives it remarkable climbing powers without increasing its size as a target. It is always an advantage to a fighting machine to have as small a wing area as possible, for, besides being able to manoeuvre more quickly, it furnishes a smaller target to the enemy's gunners. The triplane can mount rapidly into the upper air, so as to command a strategic position above the airplanes of the foe, while to those attempting to fire upon it from above or below, its three wings do not present any larger surface than two of the biplane or the one of the monoplane.

The Curtiss factory has been at work for several years on the problem of the small fast fighter. Its first effort was a biplane whose top wing span was only 20 feet. In a test flight by Victor Carlstrom at Sheepshead Bay Speedway, New York City, its unusual performances amazed the spectators. With startling swiftness the pilot mounted into the blue, manoeuvred his little biplane with the agility of an acrobat, gave excellent tests of speed, and descended. Reducing the speed of his motor but not cutting it off entirely, he allowed the little airplane to skim slowly along the ground. Then, alighting, he took hold of the fuselage close to the tail, and steered the diminutive craft to a suitable spot from which to make another flight. With the motor still running, and much to the surprise of the onlookers, he stepped in once more, put on full power and was off.

This little airplane was nicknamed the *Curtiss Baby Speed Scout*. In one interesting respect it was different from the *Nieuport*, whose upper plane had to be cut away to increase the pilot's range of vision. In the Curtiss machine the pilot sits just behind the planes, so that he can see above and on all sides with the greatest ease. As a protection in battle his seat and the front portion of the fuselage are surrounded with thin steel, and the pilot sits close to the floor, so that he does not offer a very good target to the enemy's stray bullets. The "baby" biplane is fitted with a standard V-type motor of about 100 horsepower, and it carries fuel for a run of two and one-half hours.

The British have done some very fine work in developing airplanes of the speed scout type. Their fighting machines flew over the lines and downed the German planes in goodly numbers. Among those which earned fame for their achievements are the *Bristol Scout*, familiarly known as the "bullet," one of the fastest of the military airplanes; and the *Vickers Scout*, another of the swift eagles that helped to maintain Allied supremacy in the clouds. One of the interesting features of the Vickers scout is the high "stagger" of its planes. By this we mean that the upper plane has been set far forward, so that it appears to overhang the lower. Quite recently another British scout machine, a *Sopwith triplane*, was flown by the British Royal Flying Corps, and it made a splendid record of victories over the lines.

In a crack regiment of veteran fighters it is hard to pick out the men who might be said to be the "best soldiers." Each man excels in some individual way, and in just the right situation might prove to be the leader of his fellows. So it is bound to be with the long list of valiant little fighting planes that took up the cudgels against the Hun. No short summary can do justice to them all. There are the *Avro*, for instance, and the *De Havilland Scout Biplane* of the British, as well as a biplane of the *Sopwith* type; while the list is almost endless of British and French machines bearing such well known names as *Farman*, *Caudron*, *Dorand*, *Moineau*, *Morane-Saulnier*, etc.

But whatever the particular features of these scout machines, their armament is generally about the same. It usually consists of a machine gun operated by the pilot and firing across the propeller. The pilot directs the nose of his machine straight at the enemy and lets go a rain of bullets.

Fighting tactics are the subject of the most intense study on the part of every pilot of a scout machine. Often he has his pet system of downing the enemy. Immelmann, the famous German aviator, liked to get high in the upper air and there await the approach of a "victim," when he could dive straight down upon the unsuspecting airplane and open fire. Every pilot aims to surprise his enemy. To do so he must often perform startling aerial tricks, looping the loop to come up under the tail of the other machine, swooping down from above, or firing from behind while the tail of the enemy machine shields him as he gets in his fatal shot. The aviator learns to hide behind a cloud, to take advantage of blinding sunlight or any other natural condition in order to take the opposing airplane unawares.

It is for this reason that machines are needed which combine speed, exceptional climbing powers, and quick manoeuvring ability. Not only must they be able to practise all manner of tricks and stunts in order to surprise the foe, but it is quite as important that they be able to move rapidly on their own account, for a slow moving airplane is more liable to surprise than one which is swift in flight and able to alter its course repeatedly or else climb out of danger's way.

How important the agility of these little fighting planes is they are apt themselves to discover when one of their number meets a big reconnaissance machine of the enemy. The latter, with its big guns, is a formidable object, and could easily get the better of the lightly built combat plane, if it were not for the fact that its weight and slow speed make it unmanageable. The smaller machine drops down upon the big fellow suddenly, firing a volley at its gunners. If he kills them well and good, but if not he must perform his cleverest aerial stunts to get out of their way, or he will soon be a mere ball of fire shooting earthward. Fortunately, he is quick, and a few acrobatic turns save him from threatening disaster.

Before the present type of reconnaissance craft, bristling with machine guns had been developed, it was customary for the airplane doing photographic work, artillery "spotting" and similar duties to rely for its protection on a number of speed scouts, who flew above and around it and escorted it upon its mission. Today the airplane that is used for general service duties over the lines is a dreadnaught of the air, and although it may still take along with it on its errands a few scouts to give battle to the faster airplanes of the enemy, yet on the whole it is self-reliant and has little to fear.

To these slower-flying, larger general service machines are entrusted some of the gravest duties of war. They are the eyes of the army, whether they act for the heads of staff, flying out over the territory of the foe with their trained observers and bringing back accurate information about the movements of troops, whether they help in "spotting" targets for the gunners, or whether during an actual engagement they act as aerial spectators and messengers, helping to coordinate the efforts of the various bodies of troops.

From the beginning of hostilities Germany strove to overwhelm the French in the air and prevent their airplanes from performing these necessary duties. France was at first but poorly equipped with machines of the type so sorely needed to maintain her air supremacy. By the skill and bravery of her airmen she managed to hold out, however, and the Huns were disappointed in never accomplishing their purpose of putting out her eyes. Her engineers were in the long run much more clever than those of Germany, and by the early part of 1915 they had ready a number of superior machines for reconnaissance and bombing. For the most part they were big *Caudrons* and *Farmans*, well armed and a good match for the German maid-of-all-work biplanes. And there were large *Voisin* biplanes, suitable for photographic work or for bombing. They were used extensively by French, British, Belgians and Italians. The *Voisin*, as in its very earliest models, is still easily recognizable by its curious tail resembling a box-kite, placed at the end of a projecting framework of four long beams or outriggers. It is a pusher type of airplane, with its propeller at the stern instead of at the bow.

Larger and more formidable grow the "aerial destroyers." Today among the super-dreadnaughts of the sky may be numbered the big biplanes bearing the names of *Moineau, Breguet-Michelin, Voisin-Peugeot,* and *Farman.* Heavily armed with machine guns they rendered valuable service to the Allies in many capacities, and they were the efficient answer to the struggle of the Hun for aerial supremacy. When in the Spring of 1918 the Germans launched their tremendous offensive at the Allies, the latter were well informed in advance of their intentions, thanks to these powerful reconnaissance planes. Swooping down close to the German lines in defiance of anti-aircraft guns and fighting machines alike, they had daily looked on at the massing of troops, the bringing up of re-enforcements for the drive, and the piling up of ammunition supplies. In spite of every effort of the enemy to make their mission an intolerable one and to prevent them from spying upon preparations for the offensive, they had succeeded in bringing back to Allied commanders accurate and detailed information. By their aid the Allies knew at what points to expect the heaviest blows, and there they collected their re-enforcements. Thus the nations lined up against the Hun were ready when the blow came, and they were able to check the tremendous onslaught by their land and air forces. What they really lacked perhaps, was not "eyes," to discover what the Germans were plotting, but a large enough number of small fighting machines to keep the enemy reconnaissance craft from spying upon their own preparations; and a large enough number of huge bombing planes to have completely interfered with the German efforts to mass re-enforcements and ammunition for the push.

In the long run it is perhaps the bombing plane that represents the greatest saving in human life in time of war. An army may be well equipped with reconnaissance machines and speed scouts, so that it may keep in closest touch with every move of the enemy. But unless it is able to interfere with those moves *before* they reach the proportions of a direct and staggering blow, then the best it can do is to concentrate its own troops and supplies in readiness to meet the blow when it does fall. That means that hundreds of thousands of lives of infantrymen will be sacrificed in checking the waves of enemy troops.

The Allies discovered a far better and more economical way of winning the war than this, and in the last year of the war they strained every nerve to put it into actual operation. It was this: to search out every military base of the enemy, every munition dump, nest of guns, supply train or troop train and drop bombs upon it. Two men in a bombing machine can attack and perhaps destroy a force which, if allowed to reach the front lines, would have to be met by several thousand infantrymen. Two men in a bombing machine can destroy at a single blow the ammunition which, if it had reached the front, might have swept out a regiment.

That is why so much thought and genius has been expended upon the bombing plane. The day bomber becomes the right arm of the infantry, flying low over the lines, attacking troops and striking terror to the heart of the enemy as the huge Allied tanks did when they first started on their irresistible slow walk across trenches, troops, buildings and every manner of obstruction. The big bomber—particularly if the fighting machines have cleared the way ahead of it—is something like that: it is an invincible weapon of destruction, wiping out whole bodies of the foe at every stroke, like a giant sweeping the pigmies of earth ahead of him with his strong right arm.

The big dreadnaughts of the air like the *Moineau*, the *Voisin-Peugeot*, the *Breguet*, and the *Farman*, become, when a bombing apparatus is substituted for their camera and radio, very efficient day bombers. There is a long list of others: as for instance, the British *Avro*, *Handley-Page* and *Sopwith* machines and the French *Caudron*, *Dorand* and *Letord*.

Many of these big bombing planes were designed for long distance work either by day or by night, and so they have been made enormous weight-lifters, with large supporting surfaces, two or more

A Handley Page Machine tuning up for a flight

engines, and equipped with a fuel supply sufficient for long runs. In order to carry their engines conveniently they very often have more than one fuselage. Sometimes the pilot sits in a large fuselage in the centre, while the motors are carried in two smaller cars or bodies called "nacelles" at either side. The British *Avro*, for instance, is a huge biplane with three fuselages and two rotary engines. Its upper and lower wings are equal in span, and it can readily be distinguished from the British *Handley-Page*, whose upper wing has a large overhang. The *Handley-Page* is one of the largest machines built. It carries its two 12-cylinder Rolls-Royce engines in small nacelles between the main planes, and it can be recognized by these and its biplane tail.

The *Caudron* is another big twin-motored machine, used by French, British and Italians. Its two rotary engines are fixed in small nacelles between the planes, while the pilot rides in a centre fuselage. Somewhat after the manner of the Voisin, it carries its tail at the end of a projecting framework of four long beams, the lower two of which act also as landing skids.

America, like the rest of the nations, has had her secret ambition to try her hand at building bombing machines. In 1918 the designs for the *Handley-Page* bomber were brought to this country, and on July 6th the first American built *Handley-Page* bomber was successfully launched into the air at Elizabeth, New Jersey. The huge machine was christened the *Langley* after one of the early experimenters with the heavier-than-air machine. It had a wing span of 100 feet, and a central fuselage 63 feet long. Small armoured nacelles at either side of the fuselage carried its two 400 horsepower Liberty motors, each turning a separate propeller. Laden with its full supply of bombs, its two Browning machine guns and fuel for a long run, this giant of the skies weighs about 9,000 pounds. Our country has instituted a program of construction for these super-dreadnaughts, and before long they will form an enormous aerial weapon in the hands of our airmen. For America, still practically a novice at airplane construction on a large scale, to be able to produce in her factories the largest and most complicated of the foreign types, speaks well for her determination and resourcefulness.

The Allied nations have vied with each other in their efforts to produce the king among bombing planes. The Italians have undoubtedly carried away the prize. Their *Caproni* triplane is among the largest in the world. The details of its construction were kept secret, as it was one of the most dreaded weapons of the Allies. Three powerful Fiat

THE LAUNCHING OF A LANGLEY, A GIANT BOMBING AIRPLANE

motors drive it at a speed of about 80 miles an hour. With its five tons of bombs, destined for important objectives in the land of the enemy, it is an object to inspire awe.

The *Caproni* makers have long been known for their large bombing machines. Their three bombers, including a smaller triplane and a biplane, headed the list of their fellows at the front. In October, 1917 a *Caproni* biplane was demonstrated in America, covering a distance of almost 400 miles in about 4½ hours. It started its journey from Norfolk and landed at the Mineola aviation field, with seven passengers on board. *Caproni* bombing airplanes carried out many historic raids, among them being that on the famous Austrian base at Pola. To reach it the Italian aviators had to travel by night across the Adriatic, and they carried out their pre-arranged plan of attack with the utmost punctuality, in spite of the tremendous difficulties that loomed along their path. Two squadrons of machines left the aerodrome, the first sometime before the second, and each airplane following its fellows at a considerable distance. At 11 o'clock at night the first of the bombers flew over Pola and discharged its rain of high explosives. In rapid succession the others followed, letting go their missiles upon stores of ammunition, docks, and every object of military importance. In order to aid them in picking out their targets the raiders made use of an ingenious contrivance which so amazed and stupefied the Austrians that for a while they failed to make any attempt to shoot down the Italian planes with their anti-aircraft guns. It was a huge parachute, to which had been attached a powerful chemical light. Descending slowly the terrifying object hung as it seemed suspended in mid-air, lighting the way for the raiding machines, who took advantage of the terror of the Austrians to drop 14 tons of high-explosives and make their escape unharmed.

The tremendous *Caproni* triplane is almost impregnable. Its enemies have little chance of downing it, for it can fly even when one of its planes has been severely damaged, and with its three powerful motors it is practically immune from any engine trouble, as in case of an accident or injury to one motor the other two, or for that matter, one of them, will carry it safely home. With the great stability given it by its three supporting surfaces it can go through the stormiest weather without the slightest need for fear. Once its load of bombs has been discharged, it can rise to 7,000 feet to escape from its pursuers.

The story is told of an Italian aviator, Major Salomone, who escaped in a *Caproni* when attacked after a bombing expedition by a squadron of Austrian speed scouts. His enemies succeeded in wrecking one of the big engines by their gun fire, and in killing two of his gunners and a pilot. He himself was severely wounded, but keeping control of his machine he managed to reach home safely by the power of the remaining two engines.

The triplane is by far the best type for these giant raiders that fly by night. Their requirements are great lifting power and great stability, and these, the triplane with its extra lifting surface, best fulfils. Equipped with two or three engines so that its power-plant can be absolutely relied upon in every emergency, with accurate bomb-sighting instruments and with a compass, searchlight and other apparatus necessary for travelling by night, the triplane can be depended upon to inflict gigantic blows upon enemy bases.

The British have a big bombing triplane that was heard from in Germany: the *Sopwith*. Its three planes are equal in span, and have only one strut at each side of the fuselage, with the wiring also greatly simplified, in order to reduce the head-resistance to a minimum.

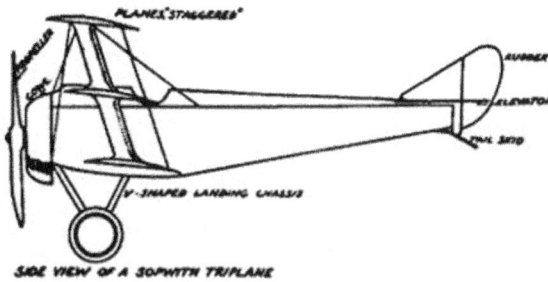

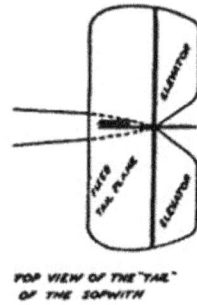

SIDE VIEW OF A SOPWITH TRIPLANE

TOP VIEW OF THE "TAIL" OF THE SOPWITH

The *Sopwith* was one of the first triplanes to be used for bombing and general service over the lines. Those at the front early in 1918 were equipped with a 110 horsepower Clerget rotary engine. A round metal hood or "cowl" surrounding the motor formed the front of the fuselage, overhanging the body slightly at the bottom in order to form an air outlet for the engine.

America has not actually developed any big bombing planes of the type of the *Sopwith*, although we have one enormous triplane,—the *Curtiss* triplane air-cruiser, built for service over the sea.

And although Russia abandoned the good cause for which she was fighting, we cannot pass over the subject of big bombing triplanes without mentioning the giant *Sikorsky*, one of the largest and most remarkable weapons of destruction that were employed in the war against the Hun.

The future will no doubt write a new and fascinating chapter in the story of the triplane. The big night bombers are being built on a large scale by all the Allied nations. Their exploits opened every great military operation, they constituted a reign of terror over the lines of the enemy, and their death-dealing blows saved countless thousands of allied troops from the need of sacrificing their lives. They could make the journey straight to the heart of the enemy's country and return, with plenty of surplus fuel. Their missiles did enormous damage to railway centres, docks, bridges, aerodromes and arsenals. Carrying bombs that weigh anywhere from 16 to 500 pounds, they spread havoc in their wake, while the silencers on their engines made them veritable spectres of the night. An illustration of their possible accomplishments was the flight of Italian machines across the Alps and to Vienna, when they dropped manifestos upon the frightened populace. Those manifestos reminded the Austrian people that only the humanity and self-respect of the allied airmen made them drop "paper bombs" on Vienna while the Germans were unloading high explosives in the midst of the civilian populations of London and Paris. It must have shown the people of Vienna what the machines of their enemies were capable of doing.

But the airplanes of war whose acquaintance we have made so hastily in this chapter were not used by the Allies for raiding or terrifying civilians. From the tiny fighting machines that carried so many of our bravest pilots to personal combat over the lines, to the enormous bombing planes used to scatter destruction and ruin among the military strongholds of the enemy, our machines were trustworthy and brave, but they were also machines of honour.

AN AMERICAN BUILT CAPRONI AIRPLANE

THIS CURTISS TRIPLANE HAS A SPEED OF ONE HUNDRED AND SIXTY MILES
AN HOUR

CHAPTER 7

German Airplanes in the World War

When we read the story of the wonderful contributions made by France, England, Italy, and America to the progress of aviation and to the romantic history of the heavier-than-air machine, we must remember that it is the story of nations which, a few short years ago, had no thought of turning the airplane into a mere weapon of destruction and desolation. It was the conquest of the air, for its own sake, that appealed to the fiery imaginations of the French, and that made them, from the day when the first Montgolfier balloon went soaring into the clouds, down to the early triumphs of the airplane in France and the great contests and meetings that followed them, ardent enthusiasts over each and every form of aerial sport. England, in spite of the fact that her sportsmen fliers were winning new triumphs daily, and in spite of the public interest that was taken from the very beginning in the advance of aviation, had, at the beginning of 1911, just *one* military airplane. America, ardent devotee of Peace, even while the World War was raging in Europe, failed to take steps to provide herself with an aerial fleet.

But when we come to Germany, the story of aviation takes an entirely different turn. The Germans as a people were never wildly enthusiastic over airplanes, for they lacked the fine sportsmanship and love of daring adventure which produced so many clever aviators in other lands. In fact, until they saw its utter inability to compete with the heavier-than-air machine as a military weapon, they confined themselves almost entirely to the construction of the safe and comfortable dirigible. With the possible exception of such a man as Lilienthal, the Germans took slight personal interest in the subject of human flight. It was the German government that, by lavish expenditure, and by every means known to it, encouraged experiment and progress.

The whole thought in Germany, both in the days of the dirigible and later, when the airplane had proved its superiority, was solely to develop the flying machine as an instrument of war. It was for this that she began her costly and gigantic program of Zeppelin construction, it was for this that the best engineers in the Empire were set to work designing aeronautic engines. It was not without some chagrin that the German military authorities gave up their dream of world conquest by means of the Zeppelin, and set themselves to building airplanes instead. Yet when they did, they applied to the new problem the same thoroughness, the same military precision and uniformity that had marked their earlier program. Reading of the French machines we are fascinated by the many types and patterns that the ingenious Frenchmen were able to devise. In Germany everything was carefully systematized by the government, individual designs were discouraged unless they fitted into the military scheme of things, and the airplane was produced in large numbers, like so many blackjacks, all exactly alike, to be used in striking the peaceful nations of the world.

German thoroughness went a long way in perfecting the airplane as a war instrument. When, in August 1914, her sword finally descended, she had close on to 800 machines and a thousand trained pilots, together with a small force of seaplanes and pilots. Today, according to an English authority, she has at least 20,000 aircraft of all sorts, manned by a force of 300,000 pilots, observers, and bombardiers.[1]

The first German machines to fly over French territory might well have struck terror to the hearts of the plucky French, for they were equipped with the cleverest instruments of destruction that Germany could devise. The swept-back, curved wings of these standard biplanes won them the name of *Taube* or "dove." Certainly they were not "doves of peace." They were equipped with wireless, carried cameras for reconnaissance work, had the most accurate recorders of height and speed, dependable compasses, instruments for bomb-dropping, dual control systems, so that they could be operated by either pilot or observer, and dozens of other small improvements and accessories that made them more than a match for the French machines sent up to dispute their supremacy in the air. The challenge these machines presented to the genius of the French was taken up with vigour. It was not long before they found themselves an obsolete form of aircraft in the great war in the air, and for all their inventions and improvements,

---

1. *The German Air Force in the Great War* by Georg Paul Neumann also published by Leonaur.

they were forced back into their hangars.

By the Spring of 1915, the French were soaring through the sky in fast fighting machines that made the air a very unsafe place for the plodding German" maid-of-all-work." The Germans bestirred themselves to think of some method of getting even with these unreasonable French pilots, who somehow refused to admit defeat. The machine which they sent out in answer to the *Nieuport* monoplane and others of its type was the invention of a Dutchman; it succeeded in creating quite a sensation for a while in Allied circles, until like others of its company it was superseded by French inventive genius and rendered a more or less harmless craft.

This supposedly invincible fighter was the *Fokker*. In general construction it was largely an imitation of the French Morane monoplane, but it had one entirely new feature that rendered it at the time a formidable adversary. That was what was known as a synchronized gun, firing through the propeller.

The problem had been to design a machine which could be operated by one man, who became both the pilot and the gunner. In order to do this he must necessarily be able to control the direction of his machine in flight and aim his gun at the enemy at the same time. The best way to accomplish this was to point the nose of his machine at his victim and fire straight ahead of him. But here the propeller was the great obstacle. How could he fire a gun from the bow of his machine without striking the propeller blades as they whirled swiftly about in front of him? The German *Fokker* answered that question. The machine gun with which it was equipped had its shots so synchronized, or "timed," that, impossible as it seems, they passed between the rapidly revolving propeller blades without striking them. The *Fokker* was a remarkable climber in its day, and in addition it had a simple device by which the pilot could lock the control of the elevating planes, steering only to right or to left, by means of pedals worked with his feet.

Early in 1916 this deadly weapon of aerial warfare made its appearance, and for a while the civilian population of England and France read with dismay of its conquests. Mounting high into the clouds, it would await its victim. The moment a machine of the Allies appeared beneath it, the *Fokker* turned its nose straight down and went speeding in the direction of its prey, opening fire as soon as it got within range. There was no use of the unfortunate airplane trying to escape. The *Fokker* could, by wobbling its nose slightly in spiral fashion as it descended, produce, not a straight stream of bullets ahead of it but a cone of fire from its machine gun, with the victim in the centre of the circle. Whichever way the latter turned to escape it met a curtain of bullets which could destroy it. The Allied machines could only combat it in groups of three and for a time at least it held supremacy in the skies. When itself pursued by a superior number of planes, it was quick as an acrobat, and speedy at climbing, so that it very seldom could be caught.

This was the machine in which the two famous German airmen, Immelmann and Boelke[2] performed some of their most daring exploits. It travelled at a speed of more than 100 miles per hour and could perform surprising feats with the most alarming ease.

But while the *Fokker's* début over the trenches caused the British House of Commons to debate the new peril gravely, French and British airmen sprang quickly and gaily to the challenge. Heedless of the danger, they braved the bullets of the *Fokker* in order to get a better view of its mechanism, and they soon answered it with swift and powerful machines like the British *De Havilland*. It was only a short while before the Fokker monoplane was "behind the times." Faster machines with greater climbing powers overtook it in the skies and swooped down upon it from superior altitudes, as it had swooped down upon so many of its victims. Its day of triumph at an end, it withdrew to the seclusion of its hangar, and the *Fokker biplane* replaced it in the air. This in its turn became the steed of many of Germany's star aerial performers.

Now came the days when Captain Baron von Richthofen[2] held forth in the heavens with his squadrons of variegated planes which the British airmen nicknamed "Richthofen's circus." These queerly "camouflaged" planes were German Albatroses. The *Albatros* was one of the best designed of the German airplanes, and although the first models produced were not remarkable for their speed, they were good climbers and weight-carriers and thoroughly reliable. They were later developed in two distinct types: a fast "speed scout" biplane single-seater, equipped with two machine guns both firing across the propeller; and a slower reconnaissance airplane, for general service over the lines. The latter carried both a pilot and an observer, and had two machine guns, one to be fired by each of them.

It was not long before the Allies had several captured machines of

---

2. *Richthofen & Böelcke in Their Own Words* by Manfred Freiherr von Richthofen & Oswald Böelcke *The Red Battle Flyer* by Manfred Freiherr von Richthofen & *An Aviator's Field Book* by Oswald Böelcke also published by Leonaur.

this type in their possession. An Austrian *Albatros* reconnaissance bi-plane, taken in 1916, afforded an interesting opportunity to examine what was at that time one of the very best of the enemy's planes. Its general construction did not entirely meet with the approval of expert airmen who looked it over. Its upper wing was much longer from tip to tip than the lower, producing a very great overhang. From the point of view of the pilot this had its advantage, for the shorter plane below him allowed a much better range of vision, but it undoubtedly weakened the whole structure. The biplane was exceedingly slow in flight, a great drawback even in a machine not built for fighting purposes. One curious feature was its very large fixed tail plane, to which the elevating plane was attached; while a decided defect from a military standpoint was the entirely unprotected position of the pilot and the observer.

Obviously the Germans had not yet solved the problem of air supremacy to their complete satisfaction. But their engineers and designers were busy thinking it over, and soon they had ready a number of swifter airplanes, foremost among which were probably the *Aviatik* and the *Halberstadt*. The *Aviatik* made great claims of superior accomplishments over the front lines. German pilots boasted that it had a "ceiling"(a climbing capacity) of almost 16,000 feet with pilot, observer and a fuel supply. This was over 4,000 feet greater altitude than that which any other Allied or enemy machine could reach under similar conditions. The machine had an upper wing span of 40 feet, 8 inches, while its lower wing measured 35 feet, 5 inches from tip to tip. It had a strong armour of steel tubing surrounding the compartment or "cockpit" which held the seats of the pilot and observer.

The *Aviatik* was an efficient bombing biplane of its day, although larger and more powerful machines have since come into the field to supersede it. It was fitted with metal bomb-launching tubes at either side of the bow, and the bombs were released by pulling a cable connected with the releasing trigger. The *Aviatik* was armed in addition with rotating machine guns, able to fire in any direction in an aerial battle.

The *Halberstadt* was a swift fighting machine or speed scout, which made its appearance in the third year of the war and proved efficient and reliable. This and the combat planes that followed it showed greater and greater speed until by 1917 the scout machines were flying at 150 miles per hour and climbing to altitudes as high as 22,000 feet.

It was the bombing plane, however, that appealed most strongly to the German mind as an instrument of destruction. Tired, perhaps, of their efforts to produce a fighting machine which should be without its match in aerial warfare, they focussed their attention about this time upon the bomber, which in 1917 was playing an ever more important role in the struggle for air supremacy.

Early in 1917, the flower of their creative genius took to the air. It was the *Gotha* biplane, and at the time of its début it proved one of the most difficult machines to attack and down of any of those flying for the Hun. The *Gun-tunnel Gotha* it was familiarly called, owing to the unusual means of defence against pursuers that had been devised for it.

Up to this time one of the best methods of attacking an enemy plane had been to come up suddenly and fire on it "under its tail." The gunner in the machine thus attacked could not get in a single shot at his pursuer without striking the tail planes of his own machine. The portion of an airplane which can be fired on in this way without danger of return fire is said to be its "blind spot," and it was this blind spot that sent many a well-armed and powerful airplane crashing to earth when its pursuers had succeeded in outmanoeuvring it.

The *Gun-tunnel Gotha* had practically no blind spot Its designers had constructed it with a tunnel that ran the length of the fuselage, from the cockpit, or compartment where the pilot and gunners sat, through to an opening just under the tail planes. A machine gun in the cockpit could be pointed through this tunnel and fired at the unsuspecting victim who came up back of it according to the most approved tactics. The opening of the gun tunnel was carefully "camouflaged," so that at a short distance it could not be seen by an attacking airplane, especially one which was unprepared for it.

The *Gotha* practically bristled with machine guns. One in its bow which commanded a fairly large range was operated by the forward observer, who sat in front of the pilot. A passage-way beside the pilot's seat allowed him to reach "gun-tunnel," where, stretched flat on the floor of the fuselage he operated the gun which fired out under the tail. Above him in the fuselage sat the rear gunner, and by their combined aid the *Gotha* could keep all enemy planes at a safe distance.

These, however, were merely protective measures. The Gotha's real mission was bombing, and for this it carried a bomb-releasing mechanism just in front of the pilot's seat, on the floor of the fuselage, while behind the pilot an additional supply of the death-dealing missiles were carried in racks in vertical position.

A GIANT GOTHA BOMBING PLANE BROUGHT DOWN BY THE FRENCH

GERMAN FOKKER PLANE CAPTURED BY THE FRENCH

These were the machines which flew over England [3] and France in 1917 scattering death and destruction. Against them the machines of the Allies were for a time almost powerless, for the best of their airplanes were completely outgunned by this new terror of the skies. The new German machine was given one of its first tryouts in the Balkans, where a squadron of twin-engined *Gothas* accomplished the bombing of Bucharest. Its efficiency proved, it appeared over the lines and was also used extensively by the Germans for long distance bombing operations.

The fact that the *Gothas* flew in large squadrons made them still more difficult to attack. Yet Allied airplanes went out to give them fight, and in spite of what seemed the almost complete hopelessness of the situation, they did succeed in breaking up *Gotha* formations and in downing a few of the dread machines.

Yet another German twin-winged bombing plane was ready about this time. The *Friedrichshafen* bomber was not so large as the *Gotha*, but in many points of construction it resembled it. A biplane, it had wings that tapered somewhat from the centre to the tips. The wings were strengthened by centre spars of steel tubing, which was also used in the construction of the rudder and elevators at the tail. The pilot occupied the rear seat in the cockpit and the gunner the forward seat, while a short passage-way ran between the two. Every effort had been made at camouflage. On their upper surfaces the wings were painted as nearly as possible earth colour, so that they might be indistinguishable to a machine looking down upon them from a superior altitude. On their lower surfaces they were painted pale blue, to blend with the sky and make them invisible to an enemy plane below.

The armament of this *Friedrichshafen* bomber consisted of three machine guns, one of them firing downward through a trap door in the fuselage. It was fitted with an automatic bomb-releasing apparatus, by means of which, as one bomb was released, another slipped into place.

Other bombing machines appeared in 1917, as the *A.E.G.* twin-motored tractor biplane, and the *A.G.O.* twin-bodied biplane. The Germans also began construction of huge bombing triplanes, heavily armed with machine guns. With squadrons of these, the *Gothas*, and the *Friedrichshafens*, they carried out in 1917 and 1918 an established program of bombardment. The night no longer held terrors for their

---

3. *The Only One Who Got Away* by Gunther Plüschow, a republished edition of *My Escape from Donnington Hall*, also published by Leonaur.

airmen, who had learned to fly in the darkness. They made their raiding expeditions, not only against Allied troops and military bases, but also on English and French towns, killing civilians and children and destroying property of no importance from a military point of view.

By these methods the Hun had hoped to acquire the supremacy of the air which his smaller fighting machines had not yet won for him. Fortunately the French and British had been hard at work, and in answer to the forays of the German bombing planes, squadrons of Allied planes dropped their missiles in the heart of Germany. The Allied planes, however, chose military objectives, and did not aim their blows at defenceless civilians.

Stroke for stroke, and with a little extra for good measure the Allies beat back their opponents in the air. Today some of the most remarkable raiding machines in existence, whether for night or for day work belong to France and England, while America is leaving no stone unturned to build up an air navy the equal of those by whose side she fought.

Yet the war in the air, on the Allied side, was always marked by honour, decency and humanity. The enemy repeatedly showed that not mere military gains, but the savage pleasure of bombing civilians, was a part of his air program. In March, 1918, nine squadrons of his airplanes flew over Paris and attacked the city. The raid resulted in 100 deaths, besides 79 people injured, a shocking story to go down in the record of the Hun's attempt at mastery of the air.

Mr. Baker, our American Secretary of War, was in Paris at the time when this historic raid occurred. He was holding a conference at his hotel with General Tasker H. Bliss, at the time American Chief of Staff, when the French warning siren was sounded throughout the city. The city was covered with a deep fog, that completely shielded from the view of the German machines any possible objective. But they had no intention of choosing targets for their bombs,—they let them fall at random upon Paris. For almost three hours terror reigned among the helpless civilians; then the raiders, having lost four of their number to the anti-aircraft gunners, turned and sped swiftly toward their own lines. Mr. Baker said:

> It was a revelation, of the methods inaugurated by an enemy who wages the same war against women and children as against soldiers. . . . We are sending our soldiers to Europe to fight until the world is delivered from these horrors.

London as well as Paris suffered from enemy bombing planes. Raid followed raid in the Spring of 1918, but the British had so improved their aerial defences that they were able to meet the attempted ravages of the enemy with the most powerful anti-aircraft guns, which, like a wall of fire, forbade the dread monsters to come within the limits of the metropolis. Many machines in the German squadrons never got close enough to London to bomb it, but those which did let fall their terrible explosives without aim or object, killing and maiming a large number of civilians. The British were finally forced to take the only course which could have effect with the Hun. They flew into the heart of the enemy's country and gave him a taste of his own medicine. True, they chose their objectives carefully, and the targets which they bombed were munition works, railways, factories, and camps, but for all their tempered revenge they made the foe smart beneath the stinging lash that descended, again and again, upon his back.

In answer to the aircraft program of the United States, Germany renewed her energies, and her construction of airplanes during the last year of the war was on a larger scale than ever before. Her small fighting machines, or speed scouts, include the *Fokker*, the *Halberstadt*, the *Roland*, the *Albatros*, the *Aviatik*, the *Pfalz* monoplane, the *Rumpler*, the *L. V.W.* and a number of others.

Some of these we have already seen at work. The *Roland* is one of the latest types of German two-seater scouts. Every effort has been made in it to decrease the "head resistance" by careful streamlining, reduction of the number of interplane struts, etc. Swift flying and a rapid climber, it has won for itself the title of *The German Spad*. The *Pfalz* is built either as a monoplane or as a biplane. It is a machine somewhat similar to the *Fokker*. The monoplane, however, has two machine guns, one on each side of the pilot, and firing through the propeller.

Among airplanes used by the enemy for general service duties over the lines, the *A.G.O.*, the *A.E.G.* and the *Gotha* undoubtedly take the lead. All are heavily armed with machine guns and bombs and are driven by powerful motors.

Yet for all the desperate German struggle for supremacy, her machines and her pilots did not prove the equals of the Allies in the air. The airplanes of France, England, Italy and America maintained a ceaseless vigilance over the lines, giving chase to every enemy plane or squadron of planes that made its appearance on the horizon. Our airmen showed the most dauntless courage, and they continually outwitted and outmanoeuvred the slower thinking Hun. Our speed scouts challenged his reconnaissance and bombing planes, and prevented them from performing their missions effectively; our own reconnaissance airplanes gave him a hard time of it; and our bombing machines proved themselves the strong right arm of the service—taking the place of the big guns in raining heavy explosives upon enemy troops, bombing his military bases, and making life in general most uncomfortable for the foe.

It is a far cry from those first standardized *Taubes* to the many makes and patterns of German airplanes of the present day. As the Allies met those first maids-of-all-work with a mixed company of airplanes of many and untried talents, so today, (as at time of first publication), they are meeting her efforts to imitate their own versatility in aircraft with machines which are carefully standardized in every detail. It should be an object lesson to Germany that the Allies have triumphed in each case.

CHAPTER 8

Heroes of the Air

Heroes of the air in peace times have been numerous. We already know the stories of many of the pioneers of aircraft, who risked their lives in situations involving the utmost peril. The men who, in the first frail monoplanes and biplanes attempted to fly the British Channel, or to make dangerous cross-country flights under adverse weather conditions were heroes indeed.

Yet undoubtedly the greatest exploits will be told of those heroes who, in the Great War, flew daily over the lines, meeting the aviators of the enemy in mortal combat.

Every allied nation engaged in the great conflict has her sacred roll of honour of those who fought for her in the air. Americans will never grow weary of tales of the great Lufbery; Englishmen will boast of the prowess of Bishop, McCudden and the rest of them; while Frenchmen will tell, with mingling of joy and sadness, of the immortal Guynemer, Prince of Aces.[1]

Georges Guynemer's name will always stand first on the record of the war's great flying men. His short career was a blaze of triumph against the Hun, but with many a hairbreadth escape from death and many a feat of reckless daring. Young, handsome and dashing, anxious to give his life for his beloved France, he became the adored idol of the French nation. On one occasion when he marched in a parade in Paris, the people strewed his path with flowers, and it was necessary for the police to intervene and protect him from the enraptured multitudes who pressed forward to embrace him.

Yet Guynemer came near missing the fighting altogether.

Guynemer was born on Christmas day, 1893, in the town of Com-

1. *The High Aces* by Laurence la Tourette Driggs, a republished edition of *Heroes of Aviation,* also published by Leonaur.

piègne. He grew up a tall, delicate boy, who, his friends predicted, would never live to reach maturity. Perhaps the fact that he was almost an invalid turned his attention away from the athletic sports of the other boys and gave him his intense interest in mechanics. He had one consuming ambition: to become a student in the École Polytechnique in Paris; but when by hard study he had finally prepared himself and came up for his entrance examination, the professors of the school rejected him on the ground that he might not live to finish the course. To help the lad forget his overwhelming disappointment, his parents hurried him away to a health resort at Biarritz.

He had been there a year when in August, 1914, came the news that his country had been attacked. Burning with zeal to help defend his beloved France, Guynemer offered himself again and again for enlistment in the French army. Hard pressed as that army was, its officers did not feel that they needed the sacrifice of a frail youth with one foot in the grave. Gently but firmly, the young candidate was rejected. Bitterly humiliated he went back to his life of enforced inaction; and while he saw his comrades marching forth to war, he eagerly pondered in his mind what service he could perform in the war against the invader.

At length he hit upon an idea. Since he could not become a soldier, why should he not turn his mechanical skill to some account in one of the great airplane factories where France was turning out her swift squadrons of the air. He volunteered and was accepted. In a short time he had made his presence felt, for he had received a thorough preparatory education in mechanics and was far the superior of the majority of his fellow workmen. Little by little he won the friendship and admiration of his superiors, who promoted him to the position of mechanician at one of the big military aviation fields. Now for the first time he was living among war scenes. While he performed his humble duties in the hangar he burned with ambition to pilot over the lines one of the swift French battle planes. But he hardly dared make the request that he be taught to fly, fearing the rebuff which he had received on every other occasion when he had sought to enlist.

But the officers at the aviation camp had been watching young Guynemer, and their respect for his nobility of character and high intelligence finally outweighed their fears that he might prove too delicate for the service in the air. So the happy day finally arrived when he was permitted to enlist as a student airman. In January, 1916, having completed his course of training, he flew for the first time in a swift scout plane.

From the day that he first flew out over the lines, his higher officers realized that here indeed was a master airman. In three short weeks he had won the distinction of "ace," having downed his fifth enemy machine. The secret of his success lay partly in the frail constitution which had come so near condemning him to inactivity. For the youth was fully convinced that he had not long to live, and his one idea was to die in such a way as to render the greatest possible service to his native land. Perfectly prepared to meet death when the moment came, he was scrupulously careful never to court it unnecessarily, for he realized that the longer he lived the more damage he would be able to inflict upon the enemy.

The early morning invariably found him in his hangar, going over with loving care every detail of the mechanism of his swift scout plane. Not until every portion of engine, wings, struts and stays had been tried and proved in A-1 condition, and every cartridge removed from his machine gun and carefully tested, did he climb into his pilot's seat and wing his way across the sky in search of enemy planes.

And when Guynemer encountered an enemy plane he manoeuvred to overcome it with the same care for exactness of movement. His cool-headed precision made it almost impossible to take him by surprise, while there was many a hapless machine of the enemy that he pounced upon unawares. He was an accomplished aerial acrobat, and one of his favourite tactics was to climb to a great altitude and then, pointing the nose of his plane at his prey, to suddenly swoop down at enormous speed, firing as he came.

Expert as he was, the great French aviator had a number of narrow escapes from death. In September, 1916, seeing one of his fellow aviators engaged in an unequal combat with five German *Fokkers*, he sped to the scene of the affray. Manoeuvring into a favourable position above his opponents he shot down two of them within the space of a few seconds. The remaining three *Fokkers* took to flight, but Guynemer was hot on their trail. Another of them went crashing earthward. Suddenly, as the plucky Frenchman sped on, hot on the trail of the two that were still unpunished, he was startled by the bursting of a shell just under his machine.

One of the wings of his plane had been torn completely away, and from a height of ten thousand feet in the atmosphere, he began falling rapidly. He struggled bravely with the controls but nothing could check the ever increasing speed of his plunge earthward. At an altitude of five thousand feet the airplane commenced to somersault, but the pilot was strapped in his seat. Then, as if some unseen force had intervened, the swiftness of the descent was unexpectedly checked. With speed greatly lessened the airplane came crashing to the earth, and the plucky aviator was rescued from the *débris*, unconscious but not seriously hurt by his dreadful fall. It was for this exploit that he received the rank of lieutenant, while he was decorated with the much-coveted French War Cross.

On another occasion Guynemer's machine was shot down by German shells, and came crashing to earth in No Man's Land, between the French and the German trenches. The Prussians turned their machine guns on the spot and ploughed the area with scorching fire. But the French had seen their beloved hero fall, and without a thought for the consequences the *poilus* in the trenches went "over the top" after him. Quickly they bore him back to safety, and if they left some of their comrades fallen out in that dread region, they did not count it too great a sacrifice to have redeemed their idol with their blood.

Practically every fighting nation has had not only its favourite airman but also its favourite aerial escadrille. Guynemer was the leader of the famous band of "*Cignognes*" or "Storks," into which had been gathered the pick of all the flying men of France. His historic opponent in the war in the air was the German Baron von Richthofen, whose squadrons were humorously nicknamed "Richthofen's circus" by the Allies, because of their curiously camouflaged wings. The Germans were very jealous of Guynemer's successes, and as the record of the number of machines he had downed grew, they eagerly credited Richthofen with more victories. Guynemer's final score was 54 and his enemy's much higher. Yet as a matter of fact the Frenchman had destroyed many more machines than Baron von Richthofen, for whereas the French gave no credit for planes sent to earth where no other witnesses than the pilot could testify to their destruction, the Germans were very glad to pile up a huge score for their hero, and were not by any means critical in seeking proof of a victory.

Guynemer's remarkable aerial victories made him a hero throughout the world. It was reported that in one day he had been officially credited with the destruction of four airplanes of the enemy. One of his chief ambitions was to bring down an enemy machine within the allied lines, as little damaged as possible. Such a plane gave him an opportunity to indulge his interest in the purely mechanical side of aviation. With the utmost patience he would examine it in every detail, making note of any features which he regarded as improvements on the *Nieuport* he himself flew. Such improvements would very shortly appear on his own machine. So while Guynemer flew a *Nieuport*, it was in reality a different *Nieuport* from any doing service over the lines. In its many little individual features and appliances it reflected the active, eager, painstaking mind of its famous pilot, whose mind was ever on the alert to discover the tiniest detail of mechanism which might gain for him an advantage over his adversaries.

It was on September 11, 1917, that the beloved aviator fought his last battle in the air. While in flight over Ypres he caught sight of five German *Albatros* planes, and instantly turned the nose of his machine in their direction. As he bore swiftly down upon them, a flock of enemy machines, over forty in number, suddenly made their appearance and swooped down from an enormous height above the clouds. Baron von Richthofen with his flying "circus" was among them. None of Guynemer's comrades was near enough to aid him. In the distance a group of Belgian machines came in view, rushing to his assistance, but before they had arrived at the spot the plucky French airplane was observed sinking gently to the earth, where it disappeared behind the German lines.

Guynemer's comrades cherished the hope that he had been forced to descend and had been taken prisoner by the Germans. Such an ending to a glorious career of service would perhaps not have been desired by the aviator himself. He who had used his life to such good advantage for his country had crowned his victories with death. The Germans themselves, out of respect for his memory, undertook to inform his fellow-men of his fate, and a few days later they dropped a note into the French aerodrome stating that he had been shot through the head. The German pilot who had killed him was named Wissemann, and he was an unknown aviator. When he learned that he had actually killed the great Guynemer, he wrote home to say that he need now fear no one, since he had conquered the king of them all. It was scarcely a fortnight before he was sent to his death by a devoted friend of his renowned victim.

The man who avenged the death of Guynemer was René Fonck, likewise a member of the French "*Cignognes.*" Fonck took up the championship of the air where his comrade had laid it down. He stands today as the most remarkable of all the French aviators. He has been called "the most polished aerial duellist the world has ever seen." With an official record of almost half a hundred enemy machines destroyed, he has astounded his spectators by his aerial "stunts" and the absolute accuracy of his aim. Many of Fonck's successful battles have been fought against heavy odds, quite frequently with as many as five of the enemy's airplanes opposing him. Yet with apparent ease he invariably succeeded in warding off his would-be destroyers, whilst one by one he sent them flaming to the earth.

It has been said of Fonck that in all his battles in the clouds he never received so much as a bullet hole in his machine, thanks to his unparalleled skill at manoeuvring. He made a world's record at Soissons in May, 1918, when he downed five enemy airplanes in one day. He was flying on patrol duty when he came upon three German two-seater machines, and in less than 10 seconds sent two of them flaming to earth. Later in the same day he actually succeeded in breaking up a large formation of German fighting machines, and after destroying three, sent the rest fleeing in confusion.

On another occasion Fonck made a world's record when he brought down three German planes in the brief space of 20 seconds. While in flight above the lines he came upon four big biplanes of the enemy, flying in single file, one behind the other. He quickly pounced upon the leader, and in less time than it takes to tell, had sent him crashing to the earth. The second had no chance to alter its course. Training his machine gun on it Fonck soon sent it, a mass of flames, after its fellow. The third big biplane dodged out of the line and sped out of harm's way, but the fourth was caught by the plucky Frenchman, who wheeled his machine round with startling rapidity and fired upon it before it could make good its escape.

This remarkable feat, performed in August, 1918, brought Lieutenant René Fonck's official total of victories up to sixty, and made him the premier French ace, at the age of twenty-four. In all his aerial battles he had never been wounded, passing unscathed through the most formidable encounters by reason of his unparalleled skill at manoeuvring.

Guynemer and Fonck are perhaps the two greatest names on the French roll of heroes of the air. But there were many other Frenchmen who did valiant service. Lieutenant René Dorine had an official record of 23 victories when he disappeared in May, 1917. He was nicknamed the "Unpuncturable" by his comrades, since in all his exploits above the lines his machine had only twice received a bullet hole. Lieutenant Jean Chaput had a record of 16 enemy planes destroyed, when in May, 1918, he made the great sacrifice; and there are many others, some living and some fallen in battle, who, flying for France, day after day and month after month, helped to make her cause at length a victorious one.

The "ace of aces" among British flying men of the war is Major William A. Bishop, who holds the record of 72 enemy airplanes downed. Second to him on the British list stands the name of Captain James McCudden, who had disposed of 56 of his enemies when he himself was accidentally killed. McCudden had had a most picturesque career. He joined the British army as a bugler at the age of fifteen. As a private he fought with the first Englishmen in France in 1914. His first flying experience came at Mons, when owing to the scarcity of observers he was permitted to serve in that capacity. He rapidly made good, and was soon promoted to the rank of officer. He proved himself a clever aerial gunner, and so won the opportunity to qualify as a pilot. With a fast fighting machine of his own he became a menace to the Hun, with whom he engaged in over 100 combats during his flying career, yet never himself received a wound.

Other English fliers made special records in the Great War, as Captain Philip F. Fullard, who downed 48 enemy machines; Captain Henry W. Wollett, who accounted for 28; and Lieutenants John J. Malone, Allan Wilkinson, Stanley Rosevear and Robert A. Little, all with scores of from 17 to 20. Captain Albert Ball, who was shot down by Baron von Richthofen in 1917, had an official score of 43 victories over the Hun, with the additional honour of having conquered the great German aviator Immelmann.

And now we come to the story of America's great fliers. Long before America herself had entered the World War there had arisen a valiant little company of her sons, who, remembering our ancient debt to France, had gone to fight beside her men in the war against the invader. Many of these Americans became skilful aviators and members of the squadron which the French had appropriately named the "*Lafayette Escadrille.*"[2] In 1916, three of its most distinguished fliers— Norman Price, Victor Chapman and Kiffen Rockwell—gave their lives to France. Probably the name which all Americans know best is that of Major Raoul Lufbery, till his death American "ace of aces," who flew with the *Escadrille* under the flags of both countries.

Major Lufbery's personal story is romantic as any fiction. He was a born soldier of fortune. When a very young chap he ran away from

1. *Winged Warfare* by William A. Bishop also published by Leonaur.
2. *The Story of the Lafayette Escadrille* by Georges Thenault also published by Leonaur.

home and for several years rode and tramped over Europe and part of Africa, working at anything that came to hand. After his early wanderings there followed two years of strenuous service with the U. S. regulars in the Philippines; and after that another long, aimless jaunt over Japan and China. It was in the Far East that he came by chance upon Marc Pourpe, the French aviator who was giving exhibition flights and coining money out of the enthusiasm of the Orientals. The two men became fast friends and Pourpe took Lufbery along with him on his travels. As an airplane mechanic under Pourpe's direction Lufbery found his first serious employment and also his first serious interest. He conceived a deep interest in aviation and became an apt pupil.

Then came the war, and Pourpe offered his services to France. Lufbery went along as his mechanic. It was only a few months before his friend had fallen, and Lufbery, anxious to avenge his death, sought admission to the ranks of French fliers. In 1916, after much excellent service over the lines, he became a member of the Lafayette Escadrille. The spectacular period of his career had now begun. He had soon claimed the five official victories necessary to make him an "ace," and in addition was presented with the *Croix de Guerre* for distinguished bravery in action. With his swift *Nieuport* he engaged in combat after combat, coming through by sheer cool-headedness and skill born of long experience. He was officially described by the French Government as "able, intrepid, and a veritable model for his comrades."

In November, 1917, America had the honour of claiming back her son, when he became a major in the U. S. service and commanding officer of the Lafayette Escadrille. And it was with the utmost sorrow that the American public, a little over six months later, read that our great aviator had met his death. He fell on May 19, 1918, in an attack on a German "armoured tank," which already had sent five American airplanes plunging to earth. Lufbery's official total was 17 German planes destroyed, but actually he had accounted for many more. He had been made a *chevalier* of the Legion of Honour by France, and like others of his American comrades had done much to cement the friendship between the two countries.

Another American ace who deserves the gratitude of the American people, not only because he brought down twenty-six German aircraft but because of the extraordinary inspiration of his example as a leader at the front to other American air fighters, is the present premier American ace, Captain Eddie Rickenbacker, idol of the automobile racing world before the war.

America's entrance into the war fired Rickenbacker with an ambition to get into the fighting at all costs and after an attempt to organize a squadron composed of expert auto racing men, unsuccessful because of lack of funds, he enlisted in the infantry. He became General Pershing's driver at the front and while serving in this capacity watched his chance to get into the flying end of the air service. An opportunity soon presented itself and Rickenbacker advanced rapidly. In eighteen months he had, as commanding officer, perfected the finest and most efficient flying squadron in the Allied armies, and had become America's ace of aces. His service was distinguished by untiring energy, devotion to his men and sacrifice of personal ambition in the demands of his duty as a leader, for it is a self-evident fact that had Rickenbacker been a free lance, he might easily have doubled his score of victories. He is a *chevalier* of the Legion of Honour, has received the *Croix de Guerre* with three palms, and also the Distinguished Service Cross with nine palms.

A particularly lovable figure in American aviation during the war was Edmond Genet, who fell in the Spring of 1917 while serving under the Stars and Stripes. Born in America, young Genet was descended from the first French minister to the United States. The two countries were equally dear to him. When he died, at his own request the Tri-colour and the Stars and Stripes were placed side by side over his grave, as a mark, so he said "that I died for both countries."

It would be impossible to enumerate in one short chapter all the brilliant records that were made during the war by the aviators of the allied nations. The best we can hope to do is to remember those names which stood out most prominently in the long story of victories won and sacrifices made to the cause of the world's liberty. Opposing our brave men there was, from time to time, a German flier who attained considerable renown, and who, for a time at least, baffled his opponents. Thus in the early days Immelmann and Boelke were much heard of. Each had his peculiar method of manoeuvring and fighting. Immelmann's favourite trick was to "loop the loop" in order to get out of the way of an enemy's gunfire, suddenly righting himself before the loop was finished, in order to fly back and catch the opposing airman unawares. By this "stunt" he succeeded in sending 37 Allied aviators to their deaths, before he himself was shot down by Captain Albert Ball of the British Royal Flying Corps.

Captain Boelke had a totally different method of attack from that of Immelmann. His favourite pastime was to lurk behind a cloud at a

CAPTAIN EDDIE RICKENBACKER

great altitude, until he spied an airplane of the Allies below him, when he would point the nose of his machine straight at his victim and dive for it, opening fire. In case he missed his target he never waited to give battle, but continued his descent until he had made a landing behind the German lines. According to the lenient German count, he had scored 43 victories up to the time of his death. It was an American, Captain Bonnel, in the British air service, who finally defeated and killed him in October, 1916.

Early in the war the Germans discovered that, however perfect their airplanes might become, their airmen were not the equals of those who were flying for the French and British. The German works much better under orders than where, as in aerial combat, he is required to rely entirely upon his personal initiative. The Allied airmen therefore soon claimed supremacy over the lines, and it was in order to wrest it from them that the Germans began turning over various schemes in their mind. The one which proved acceptable in the end has been credited to Captain Boelke. It was that of sending German aviators out in groups to meet the Allied fliers, each group headed by a commander. This plan at least proved much more successful than the old one of single encounter. Thus Boelke became the commander of a German squadron, which after his death passed to the leadership of Baron Max von Richthofen.

Richthofen was one of the cleverest of the enemy aviators and in time he made his squadron a formidable aerial weapon. He conceived the idea of camouflaging his planes in order to render them invisible at high altitudes. Accordingly he had all the machines under his command gaudily coloured. He presented a curious spectacle when he took to flight with his gaudily painted flock of birds and the British promptly nicknamed his squadron "Richthofen's circus." The "circus" usually consisted of about 30 fast scout machines, with every pilot a picked man. Freed from all routine duties over the lines its one object was to destroy, and so it roved up and down, appearing now here, now there, in an effort to strike terror to the hearts of British and French airmen. It took a large toll of our best fighters, although Richthofen's personal record of 78 victories was undoubtedly exaggerated.

The most effective fighters against this powerful organization were the members of the world-famous Hat-in-the-Ring Squadron commanded by Captain Eddie Rickenbacker, America's ace of aces. Day after day they went out against the boasted champions of the German Air Service and day after day they came in with German planes to

THE FIRST BAG OF MAIL CARRIED BY THE U. S. AERO MAIL SERVICE

their credit. At the close of the war they had won a greater number of victories than any other American squadron. The Hat-in-the-Ring was the first American squadron to go over the enemies' lines, the first to destroy an enemy plane and it brought down the last Hun aeroplane to fall in the war. After the signing of the armistice it was distinguished by being selected as the only fighting squadron in the forces to move into Germany with the Army of Occupation. It will doubtless go down in history as the greatest flying squadron America sent to the war.

On April 21, 1918, the "circus" was in operation over the Somme Valley, over the British lines. Several of its fighters attacked a couple of British planes unexpectedly, and quite as suddenly the whole squadron swooped down out of the blue. Other British airplanes rushed to the spot from all directions and there followed a confused battle which spread over a wide area.

One of the German planes which had been flying low came crashing to earth. When the wreckage was removed and the body of the pilot recovered he was found to be no other than the great Richthofen himself.

Thus the greatest of the German champions was downed. He was buried with military honours by the British, but the menace which he stood for had happily been destroyed.

CHAPTER 9

The Birth of an Airplane

Out in the forests of the great Northwest there stands a giant spruce tree, tall and straight and strong, whose top looks out across the gentle slopes of the Rocky Mountain foothills to the Pacific. For eight hundred years, perhaps, it has stood guard there. Great of girth, its straight trunk rising like a stately column in the forest, it is easily king of all it surveys.

Someday the woodsmen of Uncle Sam come and fell that mighty spruce. And then begins the story of its evolution, from a proud, immovable personage whose upper foliage seemed to touch the clouds, to a strong and lithesome bird who goes soaring fearlessly across the sky.

Uncle Sam has had an army of over ten thousand men in the woods of Oregon and Washington during the past year, selecting and felling spruces for airplane manufacture. Only the finest of the trees are chosen, and lumber which shows the slightest defect is instantly discarded. The great logs are sawed into long, flat beams, and are carefully examined for knots or pitch pockets or other blemishes which might impair their strength when finally they have been fashioned into airplane parts. These beams then start on their journey to the aircraft plants, where skilled labourers get to work on them. For the days of the homemade airplane have passed. It is only about fifteen years since the Wright brothers built their first crude flying machine, and, not without some misgivings, made the first trial of their handiwork. Since then airplane manufacture has made many a stride.

The flying machine of those days was largely a matter of guesswork. Nobody knew exactly what it might do when it took to the air. Nobody knew whether it would prove strong enough to bear the pilot's weight, or whether it might suddenly capsize in the air and come crashing with its burden to the earth. For the parts had been crudely fashioned by the inventor's own hands. Naturally he was very seldom a skilled cabinet maker, painter and mechanician. He knew very little about the laws of aerodynamics, about stress and strain and factors of safety. He just went ahead and did the best he could and took his chance about losing his life when his great bird took to the air.

No wonder the early fliers dreaded to set forth in even a gentle breeze! No wonder there used to be so much talk about "holes in the air" and all the other atmospheric difficulties that beset the pioneers. The wonder is that any of the early fliers ever came off alive with the fickle mounts to whom they trusted their lives.

Today the manufacture of an airplane has been reduced to the most exact of sciences. Every part is produced in large quantities by skilled workmen, and its strength is scientifically determined before it is passed on to become a member of the finished airplane. Sometimes whole factories specialize on a particular detail of the airplane. Here they make only airplane propellers; there only engines; while in this factory the wings and fuselage are produced.

Let us imagine ourselves on a visit to one of the great aircraft factories which have suddenly sprung up in the United States and become so busy with the work of turning out a huge aerial fleet. The great trees which were felled in the Northwestern woods have changed greatly in appearance since we saw them last. As a matter of fact for certain parts of the airplane they should have been allowed to lie out in the sun and rain for several years to "season," but the rush to put planes in the air has made this impossible. Instead they have been treated with a special process in order to rid the wood of its impurities. Now the big beams go to the carpenters to be fashioned into the airplane fuselage. The separate boards are carefully cut and fitted and trimmed down to perfect smoothness and symmetry. Painted and varnished the fuselage resembles a fine automobile body.

In the top or roof of the fuselage one or more circular openings have been cut. Below, almost on the floor are the seats for pilot and observer, in what are known as the cockpits. While the carpenters and cabinet makers have been busy on the fuselage, more skilled workmen still have been fashioning the airplane wings. This is one of the most difficult and delicate tasks of all. Remember that the curve of the wing determines to a large extent the speed and climbing powers of the completed airplane. The wing is built up of a number of ribs which give it the proper curve and shape. Each of these ribs must be accurately manufactured from a prescribed formula. First a piece of board is turned out which looks exactly like a cross section of a wing. But there is no need for solid wood to add to the weight of the wing, and so all over its surface the workman goes, boring out circular pieces, until only a framework remains. On its upper and lower edges a flexible strip of wood is bent down to its shape and strongly attached. The rib is now complete. A number of ribs placed in a row begin to suggest the outlines of a wing. They are connected by long beams which run from tip to tip of the wing. When these have been fastened in place the skeleton is completed and the work of the carpenters is over for a little while.

The next step is to place upon this wing skeleton its linen covering. The linen is usually cut in gores or strips which are sewed together, and then the whole piece is stretched as taut as possible upon its framework, above and below the ribs. Sometimes the seams run parallel to the ribs and are tacked down to them, but seams which run diagonally across the wing have been found more satisfactory. Of course it is practically impossible to stretch the fabric absolutely tight over the frame so that it will not sag when subjected to the heavy pressure of the air. Various methods were tried in the early days to tauten and strengthen the fabric. Today, (as at time of first publication), the covered wing is treated with a substance known as "dope," which shrinks it till it is "tight as a drum."

Dope renders the wing both air-proof and rainproof. It strengthens the fabric and makes it able to bear the terrible stresses to which it will be subjected when the airplane is racing through the sky. But it cannot be applied carelessly, and right here the skill of the very best painters is brought into play. These painters spread first two very thin coats of it over the fabric, filling up the pores so that later coats will not run through into the interior of the wing. Next two or three thicker coats are applied. After this the wing may receive several coats of varnish, while if it is a U. S. service plane it gets a final covering of white enamel, which protects the fabric from the injurious action of the sun's rays.

Now the wings and fuselage of our airplane are ready, and the rudder, the elevating surfaces and the ailerons are in course of production. They are made in the same manner as the wings, with a wooden framework over which fabric is stretched and "doped." We begin to think our big bird is almost ready to be put together, but we have forgotten two important items: the engine and the propeller.

The airplane manufacturer usually does not attempt to build his own engines or propellers. He buys his engine all ready to be installed and procures his propeller from a factory which makes this its specialty.

For the propeller is one of the most difficult parts of the airplane to produce. Above all things it must be strong, and for this reason steel has been tried in its manufacture. Curiously enough it was found that the metal propeller could not stand up under high speeds and stresses as well as one built of wood.

Many kinds of wood are used in propeller construction, and the choice depends very largely on the speed and stress—in other words on the horsepower of the engine. Sometimes a propeller is built of alternating layers of two different kinds of wood. But with high-powered engines oak is very generally employed on account of its strength.

An airplane propeller is not carved out of a single block of wood, for in this case it would not be strong enough for the difficult task it has to perform of cutting its way through the atmosphere and drawing the airplane after it. Instead it is built up of a number of thicknesses of specially seasoned wood, so arranged that the surface is formed by the cross grains of the various layers. This result is produced by first piling up a number of boards to form a block out of which the propeller can be carved. The boards are glued firmly together and then they are subjected to tremendous pressure. Now expert wood carvers begin their delicate task of turning out a propeller of a given pitch. Their work requires the utmost skill, but they succeed, until gradually the finished article begins to take form out of the crude block. A coat of varnish, a fine metal hub—and our propeller is ready to be shipped to join the wings and the fuselage and complete the manufacture of a modern airplane.

There are several other items—such as the steel landing chassis, the steering instruments and the upholstery—which we must have on hand before we are ready to commence the work of assembling. When all have been procured the happy task begins. The wings are put in place, and carefully secured by wires and supporting struts. The steering apparatus is installed, the cushioned seats are placed in the cockpits, the fuselage is mounted on the wheeled chassis, and finally when all is complete the big bird is sent out for its first test flight.

If there is any one way in which the airplane of today differs radically in its process of manufacture from the airplane of a few years ago it is in this: that it is a *tested* machine. The greatest enemy of the aviator was and always will be, not so much the bullets of an enemy as the hidden flaw in his machine's construction, which makes it "go back on him" when he least expects. The pioneer aviator built himself what he considered a "strong airplane," but when he attempted flight under weather conditions not so favourable as those on which he had counted, some untested part gave way. So in the early days there were many tragedies. Today, the airplane has become a safe mount indeed, for not only is the finished machine tried out before it is put into use, but each separate part is subjected to the most exacting series of tests. If it does not bear up under at least six times the strain it will ever be called on to endure in flight, it is rejected as unfit.

That is the reason the aviator of today dares to perform all the marvellous tricks in the air of which we read. Back of the stories of heroism and daring that have come from the battle line during the Great War, and back of the great commercial feats and enterprises that are being planned for the near future, we must not lose sight of the remarkable progress in airplane manufacture and the careful painstaking research and experiment that have resulted in greater safety in the air.

Of course it was the war that spurred everyone on to do his best in the design and construction of airplanes. Before that time England and America had made very poor showings, and France, although deeply interested in aviation, had nothing in the way of a flying machine that would not seem ancient compared with the airplanes of the present time.

America came into the field of action late, and up to the time she entered the war she had practically no airplane industry whatever. Yet when she did get in she set to work with a will, and as everyone knows she succeeded in making a real contribution to aviation in the war. Every brain that could be of service in our great country was mobilized. The automobile manufacturers did much for the cause, some surrendering their trade secrets for the good of the cause, and others turning over their large organizations to airplane construction. As a result, a recent report, (as at time of first publication), stated that there were 248 factories in the United States making planes, with over 150,000 men working on aircraft. In a single year this giant industry has sprung up, and the mechanical genius of America has been focussed upon this latest problem: the heavier-than-air machine.

It is inconceivable that our country, which can boast the invention of the airplane, should in peace times allow this great industry to wane. For a long time we slept while France was forging ahead in the design and construction of machines. The commercial uses of the airplane will be numberless, and it is bound to assume an ever more important and practical role in everyday life. America has the natural resources, and now that she has developed the tools with which to work and has trained a large body of young men to be capable pilots, she should look forward in the future to maintaining her proper place among the nations in airplane manufacture. The big bird of the sky who had his birth in America and who grew to such enormous proportions during the strenuous days of war, must not be allowed to lose his American manners when he turns to peace pursuits.

CHAPTER 10

The Training of an Aviator

It is a rocky road that leads from the obscurity of civilian life to the glory and achievement of a successful "bird-man." The man—or the boy—who elects to follow it must be possessed of brains, physical perfection, and iron grit, for he will need them all if he is to become one of the "heroes of the air." With one's feet on solid earth it is easy enough to make mistakes and profit by them, doing better the next time. The airman seldom profits by his serious blunders, for he is no longer on the scene when the experts are pointing out what error he was guilty of. The moment his machine, after a run across the ground, suddenly lifts and goes skimming off into the blue, he must depend upon himself. No friend upon the earth can shout to him any advice; his own unfailing knowledge and quick judgment must dictate in every emergency and see him through until once more he alights upon this old world.

Fortunately the war has proved that there were many young men able to do just that—depend upon themselves in situations so critical that the slightest deviation from the right course, the slightest hesitation about what to do next, would have cost them their lives, and their government a costly airplane. Such men have covered themselves with glory, and have won the love and admiration of their people. But they did not achieve their daring exploits nor make their marvellous records in the air until they had passed through a series of tests and a system of training so rigid that it might well have discouraged the most stout-hearted.

Why must the aviator be physically perfect? Just imagine for one moment some of the hardships and perils he will have to face. The higher the altitude at which he flies, the more intense becomes the cold. In some regions of the upper air temperatures as low as 80° and

90° below zero have been recorded by fliers. And rushing through the air at such speeds as 150 miles an hour produces a strain upon the lungs which only the strongest and sturdiest can endure. Nor is this all. The tiniest defect in the mechanism of the inner ear may cost the airman his life, if he undertakes night flying. If only he were required to fly in broad daylight when there were neither clouds nor darkness to obstruct his view of Old Mother Earth, he might manage to get along with a less-than-perfect ear. But at night,—on a cloudy night at that, when there are no lights on earth to guide him and no stars visible in the sky—the aviator faces some of his gravest perils.

Strange as it may seem it is often very difficult for him to tell whether his machine is in a horizontal position, whether he is flying right-side-up or is toppling over at a perilous angle. The only thing which helps him in this extremity is a slight reflex action in the inner ear which warns him of any loss of "balance." In the same way perfect vision is absolutely essential to the man who must be prepared for any sort of aerial emergency. This does not mean merely "seeing well." It means the absolute working right of the lens and muscles of the eye, their quick readjustment to normal after any series of loop-the-loops, after a nose dive or any sort of acrobatic stunt an airplane may be called on to perform.

So it goes with every one of the physical requirements laid down by the military authorities for men who would become fliers—they are not just arbitrary requirements, but are based on long experience of the demands which flying makes upon the system. In peace times the aviator may be able to get along with somewhat less than the physical perfection required of the military aviator, particularly if he takes up flying merely as a sport, for he will be able to spare himself the night flying and all the other difficult feats which have been required of the aviators in the war. But the next few years are going to see many new commercial duties opening to the airplane, and the pilots who guide these great ships of peace and industry will no doubt be chosen by just as high standards as our military aviators.

The room in which the would-be military aviator receives his physical examination has been jokingly referred to as "the Chamber of Horrors," and he reaches it after a short preliminary test of heart, lungs, and ear. As he sits side by side with his fellow applicants in the outer waiting room, he cannot help a feeling of "creepiness." At intervals a doctor appears at the door of that secret chamber and beckons another unfortunate in. He remembers all the gruesome stories he had heard of happenings in that room behind the closed door and his knees commence to shake. Gradually the minutes pass and by a supreme effort he begins to recover his nerve. Suddenly the door opens and a white faced applicant rushes out. The poor would-be aviator regrets his rashness in deciding to learn to pilot one of the big birds of the air. But it is his turn next, so, appearing as unconcerned as possible, he follows the doctor in.

He is ordered to sit down in a small chair to the back of which is attached a bracket for his head. The clamps are adjusted to hold his head firm, he is told to fix his gaze on a point ahead, and then suddenly, he commences to whirl around. Round and round he goes, ten times in 20 seconds. The chair comes abruptly to a halt. He must find that point he fixed his eyes on before starting. He struggles vainly to do so, imagining that failure means immediate rejection, but his eyeballs are turning rapidly back and forth. At last they stop, the physician calls out the number of seconds to his assistant. The same experiment is tried in an opposite direction, similar ones follow, and then the unhappy applicant braces himself for one of the most severe of all the physical tests.

His head is released from the clamp in which it has been held, and he is instructed to clench his hands upon his knees and rest his head on them. This done, the chair begins whirling once more. As it comes to a sudden halt, he is sharply ordered to raise his head. He has the impression that he is falling rapidly through space, and a dizzy "seasickness" almost overcomes him. Finally his eyeballs cease their swift gyrations. The instructor has timed them with a stop-watch. He is excused from the room, and, feeling like a man who had been through a siege of illness, he makes a dash for the open air.

If the applicant for service in the air has passed his preliminary tests successfully, he may shortly find himself at one of the government's "ground schools," where his education in airplane science begins. Actual flight is still a long way off: he must first receive some rudimentary drill in ordinary "soldiering," and next be put through an intensive course of training in a positively alarming number of studies, before he even approaches the joyful moment when he may begin to think of himself as even a fledgling aviator.

In the next few weeks he must become something of a gunner, a telegraph operator, a map-reader, a photographer and a bomber; he must make the acquaintance of the airplane engine in the most minute detail; go through a course in astronomy and one in meteorology; and learn the use of the compass and all other instruments necessary in steering an airplane along a definite course. Aerial observation forms no small part of his course of studies. Sitting in a gallery and looking down upon a large relief map whose raised hills, buildings, streams, and trenches give a very fair reproduction of the earth as it will look to him when he flies over it in a machine, he learns to pick out the objects of strategic importance, and to prepare military reports which will help the staff officers in their work of directing hostilities.

Or he may have to report the results of a mock bombardment, and thus prepare himself for the duties of the artillery "spotter." In order to be able to interpret with a fair degree of intelligence the things he will see as an aerial observer, he must know a good deal about military science and strategy himself, and this forms one of the subjects in his curriculum at the ground school. His life here is a strenuous one. He rises soon after five in the morning, and from then until lights go out for the night at 9:30 he has all too little time to call his own.

Before he is finally passed out of the ground school the cadet must prove that he understands thoroughly the principle of flight, the operation of an internal combustion engine, and the care and repair of a machine. He will be able to recognize the various types of airplanes, he will have some skill at aerial observation, and he will be able to operate an airplane camera, a bomb-dropping instrument and a range-finder, a wireless or a radio instrument. He will have been instructed in signalling with wigwag and semaphore, in the operation of a magneto, in the theory of aerial combat, and in a number of minor subjects such as sail-making, rope-splicing, etc.

Thus prepared in his "ABC's," the would-be aviator finally makes his departure for the actual flying school. Here he does not shake off dull class-room routine and launch forth upon a career of aerial adventure. Quite to the contrary his intensive training in the technical side of aviation becomes even more exacting. He takes apart and puts together again with his own hands various types of airplane engines, he practises gunnery at a moving target, he assembles an airplane out of the dismantled parts.

He does, however, have that wonderful experience, his first flight. Some fine morning he is told that the instructor will take him up, and, thoroughly bundled up for warmth in a leather jacket, woollen muffler, heavy cap, etc., with goggles and other little essentials of an aviator's dress, he climbs into the machine. He expects to acquire considerable knowledge of the science of aviation on that first flight.

As a matter of fact his mind is so completely overwhelmed by the many new sensations that come to it, that it is only a long time after that he is able to sort them out and form an accurate conception of the adventure. The roar of the motor is deafening as the big bird of the air goes taxiing across the earth. He does not realize that he has left the ground, until suddenly, looking down, he sees the solid earth receding rapidly from beneath him. Then, unexpectedly the machine gets into the "bumps" and he has a few nervous moments until finally it rights itself and goes skimming off into the blue. The sun is shining and below the earth looks peaceful and friendly.

He settles himself more comfortably in his seat and begins to enjoy his little aerial journey. Suddenly, without a second's warning, the airplane dives downward. The sickening drop leaves him a trifle paler, perhaps, and he no longer has the pleasant sensation of relaxed enjoyment. He hardly knows what to expect next, and the instructor, bent on testing his nerve takes him through stunt after stunt, climbing, turning, diving. At length the airplane glides gently to earth. A short run over the ground once more, followed by a full stop; and the young gentleman who went up a few minutes ago with a good deal of vim and self-assurance climbs out with a feeling of relief and satisfaction that his feet are once more on *terra firma*.

But do not imagine that he has lost his enthusiasm for the air. If that were the case then he would not be of the stuff of which aviators are made. At the worst reckoning he has acquired an intense ambition to someday "try it on the other fellow," and this in all probability he will do, when, in the course of time he has become an experienced and seasoned airman.

In the meantime, however, he must first accustom himself to the "feel" of the air, and next he must learn the operation and control of the airplane in flight. After a few first trips as a "passenger," he will be allowed to try his hand at steering the machine. This is done by what is called a dual control system. Instead of the single control-stick and steering-bar of the ordinary airplane, the training machine has these parts duplicated, so that any false move on the part of the student flyer may be immediately corrected by the instructor. As long as his movements are the right ones, the instructor does not interfere, but the moment he makes a mistake the control of the airplane passes out of his hands.

Gradually he becomes more and more adept at guiding the big bird through the air, and can get along nicely without any interference or correction. At each lesson he has mastered some new problem. He knows how to leave the earth at the proper angle after the first short run over the ground, and how to come down again, how to turn in the air, when to cut off the power in alighting and when to apply the brakes. He learns to listen for the rhythmic sound of the engine and to know when anything has gone wrong with it.

By far the most difficult of his problems is the art of landing. As we have already seen the speed of an airplane cannot be reduced below a certain danger line if its wings are to continue to support it in the air. This danger line varies with different types of airplanes, but in all of them the engine must be kept running at a fairly high speed or the whole structure will come crashing to the earth. To bring an airplane to earth while it is travelling at a speed of 75 miles an hour is no mean accomplishment. It must not bump down heavily upon the ground, or its landing chassis will be broken, even if no more serious accident occurs. It must settle slowly until its wheels just touch, while all the time it is moving forward at the rate of a fast express train. This is an art that requires infinite practise to acquire, but it is one of the most important feats the student airman has to learn.

However, the long wished-for day finally arrives when he can be trusted to go aloft by himself. Carefully he goes over every inch of his machine, to be sure it is in A-1 condition. He inspects the engine and tests every strut and wire, then, satisfied that it is in prime working order, he climbs into his seat. That is one of the most thrilling moments connected with his aviation training. In all other flights he has known that the errors he might make could be corrected by the trusty instructor.

Now he must rely solely upon himself. With a feeling of mastery and conquest, he goes skimming into the air. He longs to prove himself. Probably he does, and not long after he receives permission to try for an aviator's certificate. This is the certificate issued by the Aero Club of America; it does not make him a full-fledged military aviator, but it marks the completion of the first stage of his progress toward the coveted goal.

In order to acquire the aviator's certificate, the candidate must accomplish two long distance flights and one altitude flight; he must be able to cut figures of eight and to land without the slightest injury to his machine. In other words he must prove to the satisfaction of his examiners that he is able to handle an airplane skilfully, barring of course any fancy exploits in the air.

He now launches on his advanced course of training. This will require at least three months of hard work, and during that time he must learn to fly a number of different types of machines which are used in military aviation. In the meantime he may perhaps go up for examination to acquire the much-coveted "wings." But do not imagine that *they* mark the end of his education. With the aviator it is very much as with the schoolboy: when he finishes one grade or stage of his progress he passes on to a still more difficult. The man who has acquired "wings" is not immune from the most trying daily routine of studies, which include the ever important map-reading, photography, aerial gunnery and what-not.

Finally, however, there does come a day when the army aviator may be said to pass out of the elementary school of classes and instructors into the broader school of experience. Many young American aviators who served during the war can look back upon such a day with a thrill. They had then their hardest lessons to learn. The map-reading, the gunnery, the trying and tedious curriculum of the aviation school become suddenly vital issues, and the facts which were learned in the classroom have to be mastered anew by *living them* in the air. The experience of one young airman on his first real assignment goes to show how the problems which seemed so easy of solution on the ground become unexpectedly difficult when the flyer is face to face with them for the first time up there above the clouds.

Fresh from his course of training, he had been ordered to take an airplane from one government hangar to another which was close up behind the front lines. He knew his "map-reading" pretty well, but he had never made a long cross-country flight before and the ground was unfamiliar. Somewhere near his destination he made a false turn, and the first intimation that reached him of the fact that he was off his course was the appearance below him of white puffs of smoke—"cream puffs" as the airmen have jokingly nicknamed them. He realized with a start that he was over the enemy's lines and was being fired at.

Without losing any time he turned his face toward home, and this time he succeeded in spotting the lost hangar and making a safe landing. But he had learned a little lesson in following his map which no instructor could have taught him half so well.

There are many lessons like that which the airman who is new at the game must master. Gradually he becomes more and more expert and more and more self-reliant. Then, if he is of the stuff that heroes

A PHOTOGRAPH MADE TEN THOUSAND FEET IN THE AIR, SHOWING
MACHINES IN "V" FORMATION AT BOMBING PRACTICE

A GROUP OF DE HAVILLAND PLANES AT BOLLING FIELD
NEAR WASHINGTON

are made of, perhaps he may distinguish himself by his daring accomplishments in the air. The more daring and successful he appears to
be, the more certain it is that he has covered that long road of careful
preparation with exacting thoroughness.

CHAPTER 11

The Future Story of the Air

Since the days when the first man ascended into the clouds in a Montgolfier fire balloon, and since the days when the Wright brothers tried their first gliding experiments and proved that men might hope to soar with wings into the sky, many glorious chapters have been written in the story of the air.

Surely the most inspiring and significant achievement in aerial progress is the great trans-Atlantic flight made in the latter part of May, 1919, by a flying boat of the U.S. Navy. A force of fliers in three airships under Commander Towers attempted the flight from New York to Lisbon by way of Halifax and the Azores, in three "legs" or continuous flights, but on account of disastrous weather conditions, only one of these planes, the NC-4, under Lieutenant-Commander A. C. Read completed the trip successfully. The enthusiasm of the entire world was fired by this feat and it is difficult to estimate fully its epochal significance.

Simultaneous with this flight and even more daring in plan, was the attempt by an Englishman, Harry Hawker, to fly direct from St. Johns, Newfoundland, to England in a Sopwith biplane. Through an imperfect action of the water pump of his machine Hawker was forced to descend and was rescued twelve hundred miles at sea by a Danish vessel. However, the highest honour is due to this man of the air who embarked on so brave an adventure.

The next trans-Atlantic flight was made about a month after the NC-4 had blazed the air route across the ocean. This was a non-stop, record-breaking trip of Capt. John Alcock and Lieut. Arthur W. Brown—an American—in the British Vickers-Vimy land plane from St John's, Newfoundland, to Clifden on the Irish coast. These daring pilots made the distance of 1900 miles in sixteen hours—an average speed of 119 miles an hour.

Although these achievements in heavier-than-air machines were of far-reaching importance, they did not fully solve the problem of trans-Atlantic air passage. It remained for the great dirigible experiment in July to demonstrate that in all probability the lighter-than-air craft will prove more effective for this hazardous game with the elements.

On July 2 the British naval dirigible, R-34, left East Fortune, Scotland, with thirty-one men on board under command of Major G. H. Scott, and made the journey of 3200 sea miles, by way of Newfoundland and Nova Scotia, to Mineola, Long Island, in 108 hours. The fact that weather conditions during this trip were very unfavourable adds to the value of the accomplishment. The return trip was made a few days later in 75 hours.

The R-34 is indeed a mammoth of the air. At the time of its flight it was the largest aircraft in the world, having a length of 650 feet and a diameter of 78 feet. It has five cars connected by a deck below the rigid bag and is propelled by five engines of 250 H.P. each. Its maximum speed is about sixty miles an hour.

The year following the Great War will go down in history as a marvellous period in aeronautic achievement. The Atlantic was for the first time crossed by aircraft and within ten weeks of its first accomplishment two trans-Atlantic flights were made, three widely differing types of aircraft being represented.

As a matter of fact we have but begun to explore the possibilities of aerial flight. During the last few years we have been thinking of the airplane solely as an instrument of war, and for that purpose we have bent our entire energies to developing it. When all the wealth of skill we have acquired during strenuous wartimes is turned to solving the problem of making the airplane useful in times of peace, there will be new and fascinating chapters to relate.

The war has done a lot for the airplane. It has raised up a host of aircraft factories in all the large countries, with thousands of skilled workers. It has given us a splendid force of trained pilots and mechanics. It has resulted in standardized airplane parts, instead of the endless confusion of designs and makes that existed a few years ago. And instead of the old haphazard methods of production it has made the building of an airplane an exact science.

People used to be afraid of the airplane and it seemed a long road to travel to the time when it would play any important *rôle* in everyday commerce or travel. The war has resulted in making the airplane *safe*,—so safe that it is apt to win the confidence of the most timid.

Yet the airplanes that we saw and read of so frequently in war time are not likely to be those which will prove the most popular and useful in the days to come. In war one of the great aims was for *speed*. Now we can afford to sacrifice some speed to greater carrying capacity. The swift tractor biplane may possibly give way to the slower biplane of the pusher type, which has greater stability. The big triplanes, such as the Russian Sikorsky and the Italian Caproni will come into their own, and yet bigger triplanes will be built, able to carry passengers and freight on long journeys over land and sea. The three surfaces of the triplane give it great lifting powers, and on this account it will be a favourite where long trips and heavy cargoes are to be reckoned with. We may expect in the near future to see huge air-going liners of this type, fitted out with promenade decks and staterooms, and with all the conveniences of modern travel.

There is a strong probability that the airship, rather than the airplane, may prove to be the great aerial liner of tomorrow. The large airship of the Zeppelin type, travelling at greater speed than the fastest express train, and carrying a large number of passengers and a heavy cargo, is apt before long to become the deadly rival of the steamship. A voyage across the Atlantic in such an airship would be far shorter, safer and pleasanter than in the finest of the ocean vessels. Gliding along smoothly far above the water, the passengers would suffer no uncomfortable seasickness, nor would they be rocked and tumbled about when a storm arose and the waves piled up and up into mountains of water on the surface of the deep. Their craft would move forward undisturbed by the turbulent seas beneath.

We can imagine these fortunate individuals of a few years hence, leaning over the railing of their promenade deck as we ourselves might on a calm day at sea, and recalling the great discomforts that used to attend a trans-Atlantic voyage. It is amusing to think that our steamships of today will perhaps be recalled by these people of the future about as we ourselves recall the old sailing vessels that used to ply the deep a generation or so ago.

The airplane, if it is to hold its own beside the airship as a large passenger vessel, will first have to overcome a number of natural handicaps. In the first place, it is not possible to go on increasing the size of the airplane indefinitely, as is practically the case with the airship. For remember that the lighter-than-air machine *floats* in the air, and only requires its engine to drive it forward: whereas the heavier-than-air machine depends upon the speed imparted to it by its engine and propeller to keep it up in the air at all. Beyond a certain size the airplane would require engines of such enormous size and power to support it that it would be practically impossible to build and operate them. Modern invention has taught us that nothing is beyond the range of fancy, and we have seen many of the wildest dreams of yesterday fulfilled, yet it is safe to say that the airplane which would in any way approximate an ocean liner will not be built for many a year to come. In the meantime, however, we will have huge machines like the Caproni and the Sikorsky triplanes, driven by two or more motors and able to make the trans-Atlantic voyage with a number of passengers, freight and fuel for the journey.

Indeed, though for purposes of long distance travel and commerce the airplane stands a chance of being superseded by the lighter-than-air machine, there are many other important missions that it can perform in the modern world. One for which it is particularly suited is that of carrying the mail. In 1911 a Curtiss airplane flew from Nassau Boulevard, Long Island to Mineola, bearing the Hon. Frank H. Hitchcock, Postmaster General of the United States, "with a mail bag on his knees." As the machine swooped gently down over the big white circle that had been painted on the Mineola field, the Postmaster-General let fall his bag. That machine was the pioneer of a system of aerial mail which will soon reach every corner of the country. During the war a mail route was inaugurated between New York and Washington. Now, with many fast machines and trained pilots freed from war duties, a system of routes which will traverse our vast territory has been laid out.

It is for work such as this that the small, fast airplanes developed during the war may prove most successful. Travelling over 100 miles an hour, in a straight line from their starting point to their destination, they will be able to deliver the mail with a speed almost equal to that of the telegraph, and far in excess of anything that can be accomplished by the express train. For not only has the express train much less actual speed, but it must thread its way through winding valleys, go far out of its course in order to avoid some impassable mountain district, climb steep slopes or follow river beds in order to reach its destination. The airplane has no obstacles to overcome. Mountains, rivers, impenetrable jungles present no difficulty to it. It simply chooses its objective and flies to it, practically in a straight line. It can jump the Rocky Mountains and deliver mail to the western coast with the greatest ease. Regions like Alaska, where letters from the States took weeks or even months to be delivered, and to which the steamship routes were closed for a portion of the year, will be brought closer home when mails are arriving and leaving every few days.

What use can be made of the large photographing planes that have been developed during the war to such a degree of perfection? In peace times they will have many interesting duties awaiting them. The motion picture producers will no doubt employ them very widely. Flying over our country from end to end they will bring back wonderful panoramic views. They will explore the beauties of the Yukon and show us the peaks of the Rockies in all their majestic grandeur. They will be taken to other continents and sent on photographing flights into regions that have scarcely been trod by human feet, and they will bring home to us remarkable views of jungles where wild animals roam. Pictures which the motion picture man of today with his camera has often risked his life to secure, the nimble photographing plane will secure with the utmost ease.

And that suggests another possible *rôle* of the airplane in times of peace: that of exploration. As we think of Peary, pushing with his valiant party across the ice fields of the far North, struggling month after month to attain his goal, and returning to the same hard effort each time his expedition failed, we cannot help wishing for his sake that the airplane had reached its present state of development when his difficult undertaking of finding the North Pole began. Who knows but that Peary the pilot might have attained his objective many years before he did, providing of course he had had a machine of the modern type to fly in. Certainly one of the coming uses of the airplane will be that of penetrating into unknown quarters of the earth. Acting on the information which we can thus obtain we may be able to open up new stores of wealth and new territories to man.

The enormous boom that has been given to aircraft production by the war ought to have at least one happy result in peace times: it should reduce the cost of the airplane. When that is brought within the means of the average prosperous citizen, we may expect to see flying become a popular sport. The man who now sets forth on a cross country pleasure trip in his automobile, will find still greater enjoyment in a cross country flight. High above the dusty country roads, he will be able to skim happily through the blue, enjoying his isolation and able to gaze out for many miles in all directions over the beautiful panorama of the earth. The plane which he pilots will no doubt be so designed as to possess unusual stability. It will to a large extent be "fool proof." Its owner will enjoy the comfortable feeling which comes from a sense of security, and at the same time will have all the delightful sensations of an adventurer in the clouds. He will find the air at high altitudes invigorating, and so he will gain in health as he never could have done by motoring over the solid earth.

When men take to flying in large numbers no doubt we will have to have some sort of traffic regulations of the sky, but these will never need to be so strict as upon the ground, for the air is not a single track but a wide, limitless expanse, in which airplanes can fly in many directions and at many altitudes. There will never be any need of passing to the left of the machine ahead of you or signalling behind that you are slowing down; for ten chances to one you will never encounter another plane directly in your line of flight, and if you do it will be a simple matter to dive below or climb over him, continuing your journey in a higher stratum of air. There will probably be laws controlling flights over cities and communities, where an accident to the flier might endanger the lives below. What is likely to happen is that certain "highways" of the air will be established legally, extending in many directions over the country. In these directions the private airman will be permitted to fly for pleasure, while at certain intervals along the routes public landing grounds will be maintained.

Landing is still one of the most serious problems the air pilot has to face, and it is to be hoped that the aircraft builders of the near future will help him to solve this difficulty. The reason for it, as we have already seen, is that the airplane secures its buoyancy largely as a result of its speed. Wings which are large enough to support it when flying at 150 miles an hour are too small to hold it in the air when its speed is slowed down. The machine has to be landed while still moving forward at comparatively the rate of an express train, and this forward motion can only be checked after the wheels are safely on the ground. If the engine should be stopped while the airplane is still forty or fifty feet above the ground, the wings would be unable to support it and it would come crashing to the earth. But this situation of course makes matters very difficult for the airman who has not had long experience in landing his machine. He must come down on a small landing field and bring his plane to a full stop before he has crashed into the other machines which perhaps are standing about. His difficulty is added to by the fact that his propeller only works efficiently at the full speed for which it was designed. When he slows down in the air preparatory to landing, it may "slip" backward through the air, instead of driving his airplane forward at the rate necessary to support its weight. In that case he is in danger of going into a spin, from which he may not have time to recover.

For these reasons it is to be hoped that the airplane of the future will have some form of telescoping wings and of variable pitch propeller. While these improvements in construction have not been worked out practically at the present moment, there is every reason to believe that they may be before long.

But whatever structural difficulties have yet to be overcome in connection with the airplane, certain it is that the big birds which we saw so often in the sky during the war, are going to be yet numerous in peace times. As for the purely military machines, let us hope that their work is over, and that they may never be called on to fight another battle in the air. Yet if other wars should come, it is certain that they would play a still more tremendous *rôle* than they have in the present struggle. We can imagine the war of the future being fought almost entirely above the clouds. The one great contest would be for victory in the air, since the nation which succeeded in driving its enemy from the sky would have complete control of the situation on the ground.

All nations will continue to increase their aerial battalions until they possess formidable fleets, and it will be these, rather than armies or navies that will go forth to settle future disputes. It is largely to the aerial supremacy of the Allies that we have to give the credit for the winning of the present war against the Hun, and it will be by maintaining their aerial supremacy that the great nations which have taken their stand for justice and humanity will succeed in enforcing the reign of Right in the world.

Thus we see man's dream of the conquest of the air become a noble thing, while the frail-winged birds his imagination pictured to him throughout so many centuries stand ready to bear him onward and upward to still greater achievements in his struggle to make the world a better and cleaner place in which to live.